ELECTRICITY AND ELECTRONICS TECHNOLOGY KNOWLEDGE BASE

Thomas G. Sticht
Barbara A. McDonald

GOALS
Glencoe Occupational Adult Learning Series

GLENCOE
Macmillan/McGraw-Hill
New York, New York   Columbus, Ohio   Mission Hills, California   Peoria, Illinois

This program was prepared with the assistance of Chestnut Hill Enterprises, Inc.

Chapter-opening icons: Kathleen McGrail

Photo credits:

The following organizations and individuals have provided photographs for this text:

Apple Computer, Inc.: p. 36
Armstrong World Industries, Inc.: p. 70
Black & Decker: pp. 5 (A-D), 27, 45, 68
Connecticut Vehicle Inspection Program, Department of Motor Vehicles: p. 90
Duracell Inc.: pp. 26, 58, 83, 109
Hewlett-Packard Company: pp. 6 (A-B), 11, 40 (bottom), 69, 75, 78, 87, 94, 108
Independent Electrical Contractors Inc.: p. 73 (top), 98
Intel Corporation: pp. 6(C), 38 (top), 64, 66
Lanier Voice Products: pp. 37, 76
McGrail, Kathleen: p. 8
Motorola, Inc.: p. 34
Philips Lighting: p. 9
Photo Researchers, Inc.: pp. 25 (Will/Deni McIntyre), 28 (Blair Seitz), 57 (Larry Mulvehill), 59 (Hank Morgan), 65 (Robert A. Isaacs), 72 (Jerry L. Ferrara), 74 (Harvey Shaman), 96 (Blair Seitz), 104 (John Hendry, Jr.), 105 (Jim W. Grace), 106 (Tom Myers)
Radio Shack/a division of Tandy Corporation: pp. 10, 38 (bottom), 39, 40 (top), 41, 88
Sears Catalog: p. 5(E)
Sharp Electronics Corporation: p. 6(D)
Simpson Electric: pp. 82, 84, 85, 86
© Stanley Tools: p. 107
Texas Instruments Incorporated: pp. 30 (bottom), 31
Timex Corporation: p. 81
Xerox Corporation: p. 6(E)
Zenith Electronics Corporation: pp. 67, 91

Library of Congress Cataloging-In-Publication Data

Sticht, Thomas G.
Electricity and electronics technology knowledge base / Thomas G. Sticht, Barbara A. McDonald.
p. cm. — (Glencoe occupational adult learning series)
Includes index.
ISBN 0-07-061525-X
1. Electric engineering—Vocational guidance. 2. Electronics—Vocational guidance. I. McDonald, Barbara A. II. Title. III. Series.
TK159.S87 1993
621.3'023—dc20

92-5870
CIP

Electricity and Electronics Technology Knowledge Base

Send all inquiries to:

GLENCOE DIVISION
Macmillan/McGraw-Hill
936 Eastwind Drive
Westerville, Ohio 43081

ISBN 0-07-061525-X

1 2 3 4 5 6 7 8 9 0 RRD-H 99 98 97 96 95 94 93 92

TABLE OF CONTENTS

PREFACE

People used to think that when they got out of school, they could stop learning. Today, that is no longer true!

People who want careers in well-paying jobs have to use their knowledge and skills to learn every day. They have to keep up with rapid changes in technology. They must meet new demands for goods and services from customers. They have to compete for good jobs with workers from around the world.

The books in this program will help you learn how to learn. The *Electricity and Electronics Technology Knowledge Base* will give you the background you need to learn about electricity and electronics. The *Reading and Mathematics Information Processing Skills* books will teach you how to use your skills to learn new information.

When you complete these three books, you will be ready for more training in the construction trades. Then when you start your career, you will be able to learn new knowledge and skills. This way, you will always be able to keep up with changes in jobs. You will also be ready to move ahead to jobs of greater responsibility.

We wish you the very best of success in your chosen career field!

Thomas G. Sticht
Barbara A. McDonald

LIST OF FIGURES

CHAPTER 1

ELECTRICITY AND ELECTRONICS

Look at the picture on pages 2 and 3. It shows the floor plan for Electroserve, Inc. Electroserve supplies products, parts, and services for fast-food restaurants. All of Electroserve's products and services have to do with electricity and electronics. The picture shows the office and manager's area. It also shows the manufacturing area and the repair and service departments.

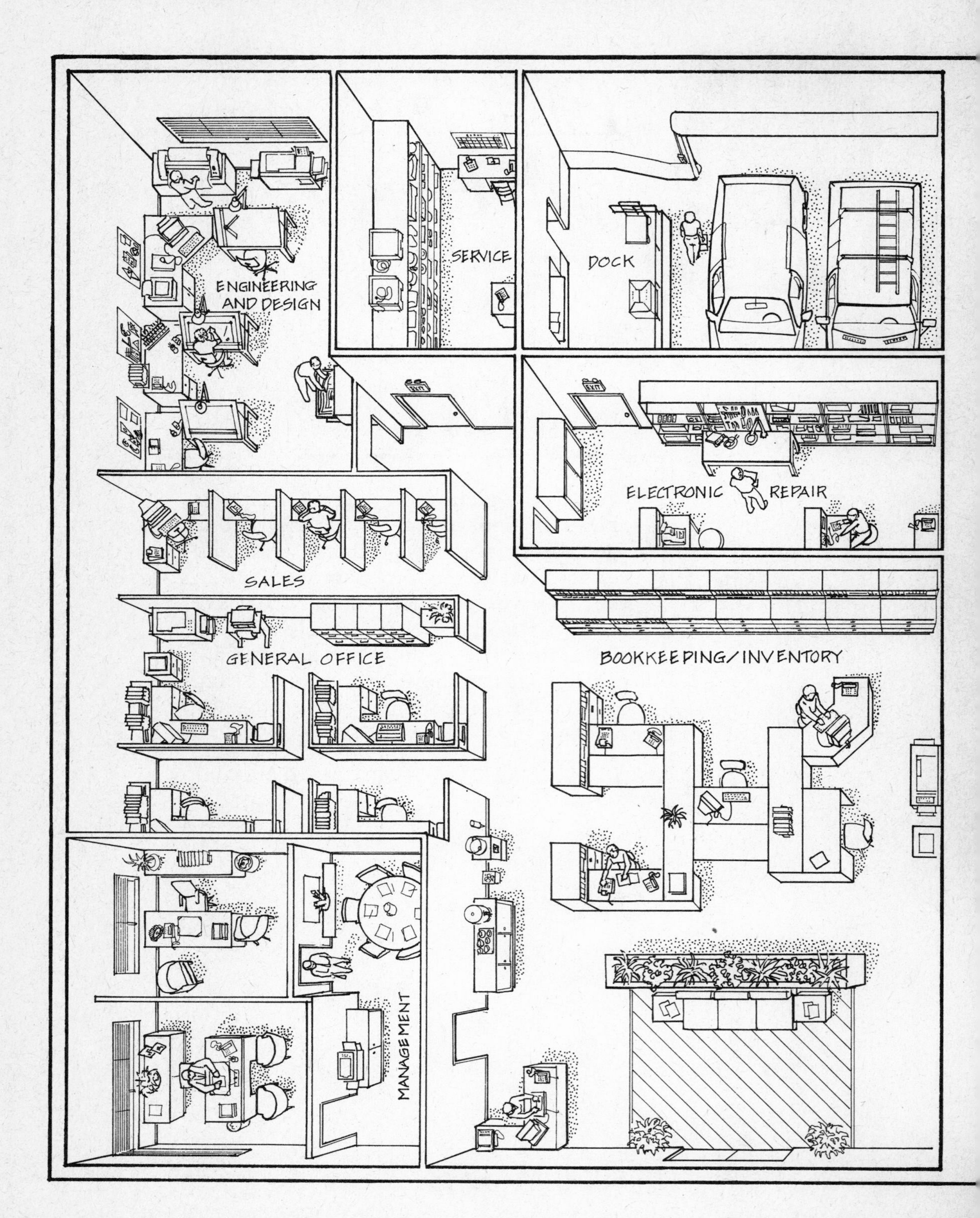
ENGINEERING AND DESIGN
SERVICE
DOCK
ELECTRONIC REPAIR
SALES
GENERAL OFFICE
BOOKKEEPING/INVENTORY
MANAGEMENT

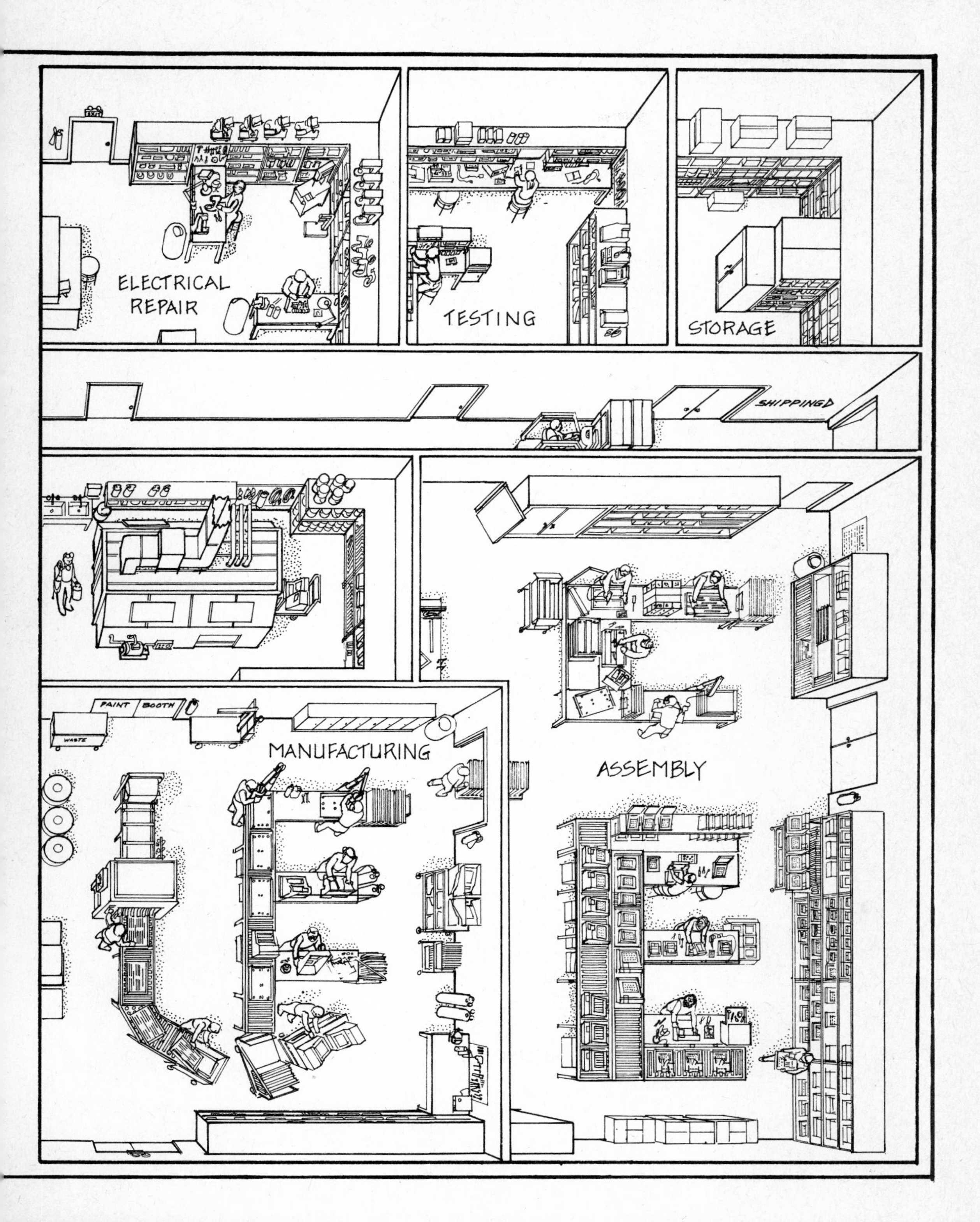
ELECTRICAL REPAIR
TESTING
STORAGE
SHIPPING
PAINT
BOOTH
WASTE
MANUFACTURING
ASSEMBLY

If you look at the table of contents, you will see that parts of this book match the areas in the picture. For example, Chapters 3 and 4 take place in the repair area. Chapter 5 takes place in the manufacturing division. Chapter 6 takes place in the assembly area. In Chapter 7, you will meet workers in the testing area.

WORKING IN ELECTRICITY AND ELECTRONICS

Almost everything we do in the world around us involves the use of **electricity.** Every time you turn on a light, use a toaster, or listen to a radio you use **electrical** energy. Workers in electricity learn to make, install, or repair **electrical circuits** or machines. Figure 1-1 shows some electrical machines. Electrical machines are used everywhere—in the home and at work.

What Is Electricity?

Electricity is the basic form of energy used in homes and businesses. Electrical workers keep power plants running. They also install, service, and repair the wires that bring electricity to all users.

What Is Electronics?

Electronic equipment, such as telephones and computers, operates by the flow of **electric** charges. This equipment is made by workers in **electronics.** Electronics is the industry that makes, installs, and repairs electronic devices. Figure 1-2 shows some electronic equipment. Electronic equipment is also found everywhere—in homes and on the job.

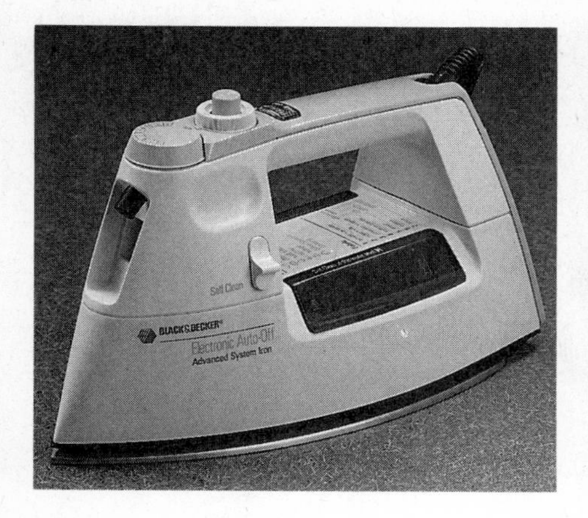

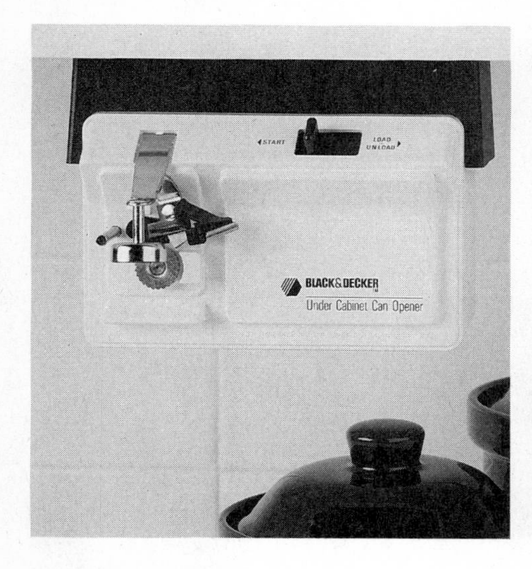

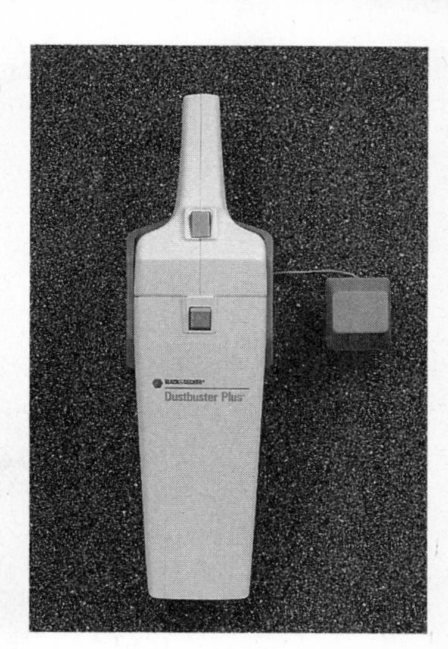

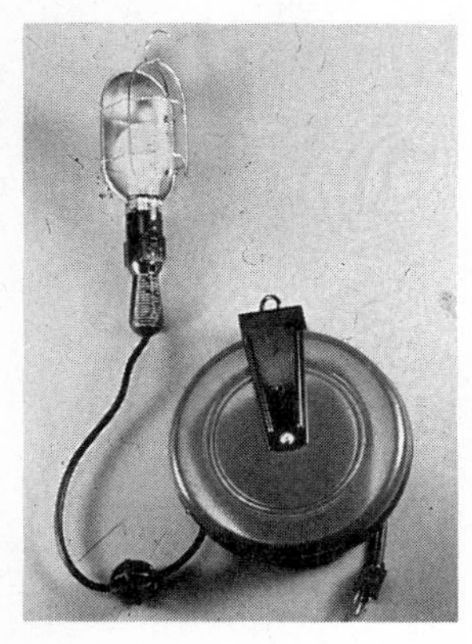

Figure 1-1 Electrical machines are used everywhere—in homes and at work.

Figure 1-2 Electronic equipment is also found everywhere—in homes and at work.

❑ JOBS IN ELECTRICITY AND ELECTRONICS

Jobs involving electricity and electronics are available in most industries. Electrical technicians can work in construction, power plants, manufacturing firms, and research companies. Electronics technicians can work anywhere electronic equipment is repaired, built, or serviced.

Electrical Technician

An electrical technician usually has some technical training. Some technicians get training while working. Others go to technical schools.

An electrical technician can work as an assistant. In construction, electricians need assistants to help with wiring. In electrical

power plants, technicians assist electrical engineers. In manufacturing, technicians perform many tasks. They can assemble and test products.

Technicians repair electrical products. They can work in small repair shops. Figure 1-3 shows a technician at work in a small repair shop.

Technicians can also work in the repair division of a manufacturing company. Figure 1-4 shows the repair area in a manufacturing company.

Electronics Technician

Electronics has become one of the largest industries in the world. Electronics technicians build, repair, install, and maintain all kinds of electronic devices.

Technicians in electronics may get their training on the job or in technical schools. They can work in manufacturing. They can also work in sales and service.

Some important parts of the electronics industry include **computers** and communications. So

much work today is dependent on computers. The need for service and sales technicians keeps growing. Figure 1-5 shows a computer showroom.

Figure 1-3 A small repair shop where the technician is repairing a blender.

Working With Computers

Electronics technicians can specialize in computers. The manufacturers of computers employ technicians in the manufacturing process. Computer repair services employ trained technicians.

Other fields to go into after getting some computer training are data processing and programming. Data pro

Figure 1-4 The repair area of a manufacturing company.

cessing is the entry of data into the computer. Programming involves entering a sequence of instructions or **programs** into the computer to help it run.

In addition, trained technicians can go into computer sales. Computer showrooms need experienced people to discuss the equipment and the software that they sell.

Working in the Energy Industry

The world runs on electric power. Electrical and electronic technicians have become very valuable to the energy industry. Not only must suppliers keep electricity flowing, they must also conserve and plan for the future. Energy needs keep on increasing.

How to use and protect the environment is a large problem in the energy field. Learning to work safely and efficiently with natural resources may be one of the most important jobs.

Figure 1-5 A computer showroom where the salespeople have to know about computers.

Professional Careers

People who decide to go further with their electrical or electronic training have boundless opportunities. The computer field keeps changing rapidly. Design engineers are always in demand.

Figure 1-6 Various testers that give instant readings.

Electrical engineers design and maintain electrical systems, such as power plants. They may also work in manufacturing.

Trained workers can move into sales and management. Many such jobs exist in manufacturing. Other management jobs are available in government.

TECHNICAL TRAINING

Electricity can be dangerous. Workers must understand how it works. They must also know all the safety rules. Students in a technician training course learn the basics of electricity. They also learn how it flows and how it can be dangerous. In this book, you will meet workers who handle electricity safely.

A technician training course also gives some hands-on experience. Depending on the course, a student may learn how to wire a house or fix a computer. Technicians find out how electrical and electronic machines work. They learn to test for and fix problems.

Technicians become familiar with **testers.** Testers are machines that test some aspect of an item. A battery tester tests whether or not the batteries are still good. Other testers test various electrical and electronic parts. Figure 1-6 shows several types of testers.

Technicians learn how to use each tester safely. Each tester has a special use. To make sure tests get accurate results, technicians have to follow exact instructions for using the testers.

CHAPTER 2

WORKING IN ELECTRICITY AND ELECTRONICS

You are a student in an electrical technician's course. Today, your instructor is taking the class to Electroserve. Your class will tour the plant and talk to employees. You will see what jobs the various workers do. You will learn about safety.

Electroserve is a company that makes both electrical and electronic products.

During your tour of the plant, you will talk to technicians and other workers. You will see how some electrical and electronic products are made and repaired.

THE OFFICE AREA

Your class enters the building. The first person you meet is John Preston. John is the office manager. He supervises the ten employees. The employees handle all office tasks. That means that they handle the payroll, inventory, order fulfillment, and billing.

Bookkeeping

John introduces you to Sal Arquilez, the bookkeeper. Sal is making up the payroll checks. He looks at a payroll register, such as the one in Figure 2-1.

Every day each worker punches a time card. The worker's hours are recorded. The time cards are collected at the end of the week. Sal figures out each worker's pay from the time card.

EMPLOYEE INFORMATION			GROSS EARNINGS			DEDUCTIONS		
NAME	MAR STAT	ALLOW	TOTAL HOURS	HRLY WAGE	GROSS PAY	FICA	FWT	TAKE-HOME PAY
Joyce Almenta	M	2		10.25				
Stephanie Jones	M	3		11.15				
David Valenti	S	3		13.25				
Jose Velez	M	1		9.75				
Lonnie Witkolski	M	4		13.00				
Sam Li	S	1		9.75				
Alberta Fay	M	3		8.60				
Herbert Knoll	M	4		10.50				

Figure 2-1 A payroll register.

Sal uses his computer to figure out each person's paycheck. The computer has a program that subtracts all the taxes and prints the checks. Payday is on Wednesday, so Sal is getting all the checks ready for signing.

Inventory Department

Next John introduces you to Katy Arnold. Katy is the inventory supervisor. Katy says, "I was once an assembler in the manufacturing division. Gradually, I worked my way up to this office job. I studied to be an electrical technician just like you. My studies helped me learn about the types of electrical and electronic devices I keep track of."

Katy also tells you, "I learned about safety. Electricity can harm or even kill you. In school, I learned to read electrical symbols. The symbols are a kind of shorthand. When I look at diagrams, I can count the number of switches a project will need by looking at a picture. This knowledge helps me in knowing about the types of products here at Electroserve."

Katy shows you the inventory system used at Electroserve. All items to be shipped have a bar code. A *bar code* is a symbol that can be read by electronic machines. Figure 2-2 shows a bar code.

Every Electroserve product has a bar code. This code is read by a device in the shipping department. The information goes in to Katy's computer. She can then keep exact track of all items shipped. The bar code and the computers form the basis of the inventory *system* at Electroserve.

Figure 2-2 A bar code.

INVENTORY REPORT		
DATE: June 1, 199X **ITEM**	**STOCK ON HAND**	**TO ORDER**
Printed Circuit Boards		
F-23X	200	______
K-150	40	______
M-17G	20	______
R-12	95	______
M-18H	20	______
Microcircuits		
P-11	720	______
XL33	90	______
ML27	50	______
R449B	1000	______
A-17	75	______

Figure 2-3 An inventory report.

A system is an organized way of getting something done. It includes a *process* that has steps. The shipping clerks make sure each product's bar code is run over the electronic reader. Then the inventory clerks have the information in their computer. Figure 2-3 shows an inventory report from a computer.

If you were to work in a fast-food restaurant, you would learn the *process* for making and selling a hamburger. One worker gets the meat and buns from the storage room. A second person cooks the meat and toasts the buns. A third person puts the sandwich together, wraps it, and puts it on a warming shelf. A fourth person sells the sandwich and collects the money.

Electroserve has systems for most of the areas in the company. As you go through the company, you will see the manufacturing system, design system, repair systems, and service systems.

You meet several secretaries. They take care of all the letters to customers. They schedule appointments for the salespeople and managers. They also perform many other office tasks, such as filing and answering the telephone.

Sales

Your class meets several salespeople. A couple of them say that they also started out as technicians. Once they learned enough about Electroserve's products, they pushed for a promotion to sales. Salespeople keep the orders coming in so that Electroserve can continue to do business.

Engineering and Design

Next, you go into the engineering and design part of the office. You meet Joan Montoya, an electrical engineer. Joan tells you, "Electricity and electronics are fast-moving industries. Electronic parts are used in so many products today. And more products are on the way. Being a design engineer is hard work, but it is fun. It feels like being part of the future."

Electroserve's major client is Quicky's. Quicky's is a fast-food chain. Electroserve manufactures, installs, and repairs various electrical and electronic equipment for Quicky's.

Joan shows you a product that she is developing for Quicky's. It will take the customer's order at the drive-thru by voice control. It will then flash the order on a board in the cooking area. The system is a year or two away from completion. Quicky's is anxious to get going on it. The company wants to keep its reputation for fast drive-thru service, and the system will help.

Joan shows you some technical drawings. These drawings show some basic designs. They show how the circuits will be connected. Figure 2-4 shows a drawing of a circuit.

The designers and engineers make up drawings for each item Electroserve sells. The drawings show how to make and repair an item or part of an item.

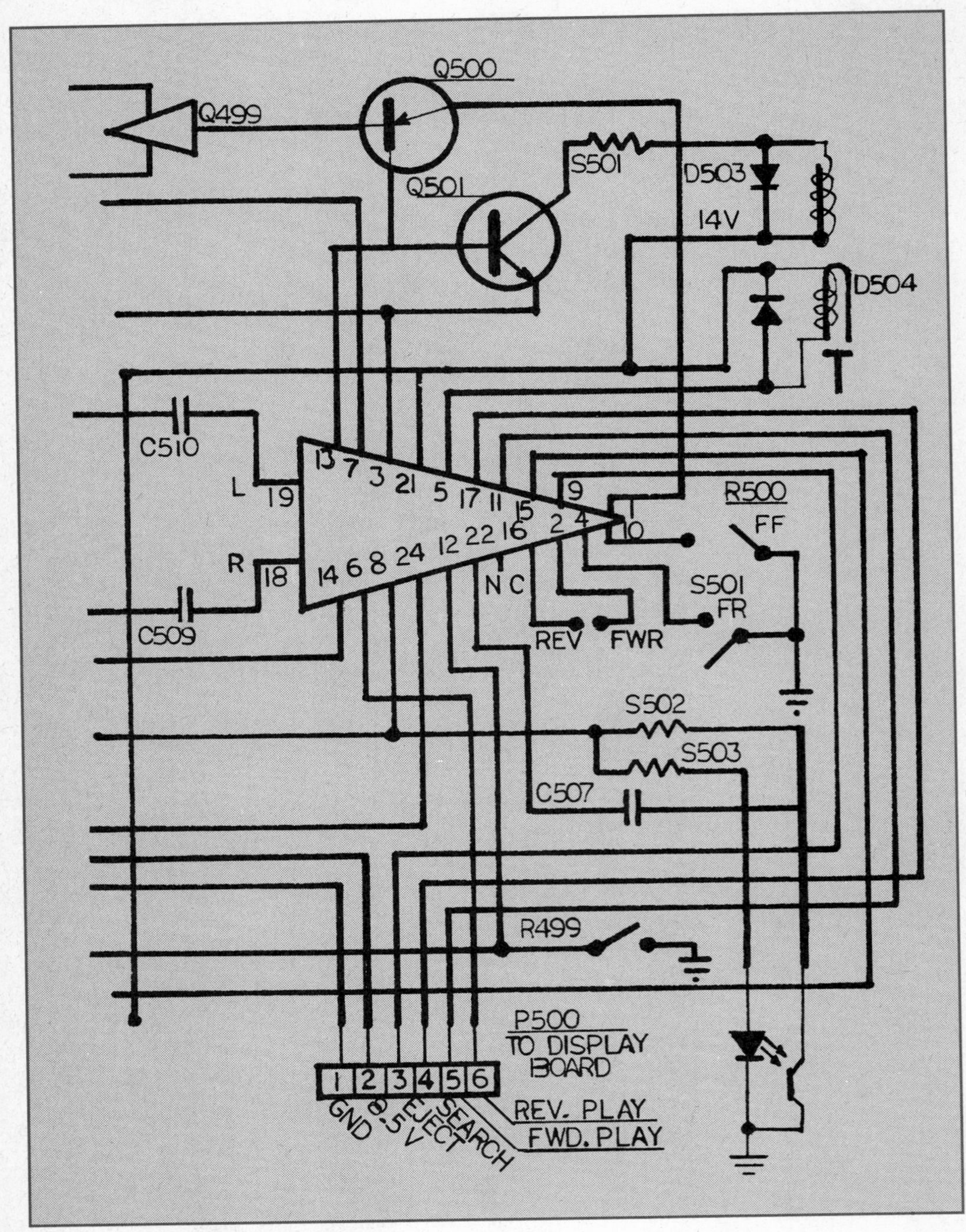

Figure 2-4 A circuit drawing.

❑ MANAGERS

Your class moves into the managers' area. This area includes a conference room and offices for all the managers. The managers supervise all the departments of the company. They also do the planning for the future of the company.

You meet Sam Johnson, who is in charge of developing new products. Sam says, "Before you tour the plant, I will describe some products you will see.

"Electroserve has been in business for 20 years. During that time, our business has changed greatly. We continue to serve our old clients. We supply them with electrical items, such as safety lights. But now, we mainly concentrate on

Figure 2-5 A register pad used in a fast-food restaurant.

electronic items, such as the drive-thru system that Joan is designing.

"Electroserve sets up the order pads for restaurant cash registers," Sam continues. "We also install all the electrical and electronic systems whenever a new Quicky's opens. Later, we service what we sell." Figure 2-5 shows a register pad on a cash register being used in a restaurant.

Throughout your day at Electroserve, you will meet workers in repair, manufacturing, and service. All the departments work together to meet customers' needs. Everybody's job depends on the company working together.

CHAPTER 3

REPAIRING ELECTRICAL ITEMS

Your class enters the repair department at Electroserve. You meet Lonnie Witkolski, the repair supervisor. Lonnie tells you that repair is divided into two areas. The first area is electrical repair. The second area is electronic repair. You will go to the electronic repair area after you learn about electrical repair.

In the electrical repair shop, you see several familiar restaurant items—toasters,
waffle irons, safety lights, and griddles. All of these items are in the shop for repair.

Lonnie introduces you to repair technician José Velez. He is repairing a safety light for a Quicky's. Figure 3-1 shows a safety light with the word *exit* on the box.

José tells you about the safety light and how it works. He explains, "When I was in school, I learned about electricity. First, we were taught what it was. Then, we studied how electrical devices worked. Because I learned these things, I understand why and how to make electrical repairs."

ELECTRICITY

José learned about electricity. **Electricity** is **energy.** It is energy that comes from the movement of tiny particles called **electrons.** Everything—from air to wood—is made up of **atoms.** Even people are made up of atoms. Electrons move

Figure 3-1 A safety light showing the word exit on it.

Figure 3-2 A model of an atom.

inside atoms. Also inside atoms are **protons** and **neutrons.** Figure 3-2 shows a model of an atom with electrons, protons, and neutrons inside it.

Electric Charge

Have you ever heard the phrase "opposites attract?" The particles inside atoms have an **electric charge,** a quality that makes them **positive** or **negative.** An electric charge is also called a **charge.** Positive and negative are opposites. That means that the positive and negative particles attract each other.

Electrons have a negative charge. Protons have a positive charge. In Figure 3-2, the protons have a plus sign (+) inside them. The electrons have a minus sign (-). These signs are a short way of indicating positive and negative.

Look at Figure 3-2 again, and note that the number of electrons and the number of protons is exactly equal. This means that the total electric charge is zero, because each positive has a negative to cancel itself out.

Sometimes the number of electrons is more than the number of protons. The "extra" electrons are called **free electrons.** Electricity is the movement of free electrons.

Current

The quantity of electrons passing through a point is called the **current.** You can compare this current to water flowing through a pipe. Scientists use the letter **I** to stand for **current.**

Sometimes, the current has a **short circuit**. This problem occurs when the current in a machine *runs around* the parts that it is supposed to *go through*. A short circuit can damage equipment and must be repaired.

The current is measured in **amperes,** commonly called **amps**. Just one ampere is 6,250,000,000,000,000,000 electrons passing through a point in one second. Imagine how small an electron is!

Another important measure tells how much work an electric device can do. **Voltage (V)** tells how much **potential energy** a particular device has. A 1/2-volt battery can light a penlight but a big flashlight might need a 9-volt battery.

Static Electricity

Figure 3-3 shows static electricity in action. The girl is standing next to a static electricity generator. Her hair is standing on end. It seems to be trying to reach the generator. The generator has a negative charge. The hair has a positive charge. And the two opposites attract. This kind of electricity is called **static electricity.**

The example of the hair and the generator shows what a strong force an electric charge is. You are studying electricity and electronics. You will learn about the movement of electrons. You will learn how to use electrical energy to do different kinds of work.

Energy and Power

Electricity is one form of energy. Energy is the capacity to do work. Work comes from different forms of energy. When you pull a sled, you use *mechanical* energy. When you cook on a gas stove, you use *heat* energy. When you turn on a light, the bulb changes *electrical* energy into light energy.

Figure 3-3 Static electricity shown in the hair of a girl standing next to a static electricity generator.

Light bulbs usually have a measure that indicates how much power the bulb has. A bulb's power is measured in **watts.** A 40-watt bulb has less lighting power than a 100-watt bulb.

Another measure of power, **horsepower,** gives the units of power in an engine. Each unit equals 745.7 watts.

Scientists and inventors use electricity. They have made thousands of machines that use, change, and control electricity. At Electroserve, you see electrical equipment being manufactured and repaired.

ELECTRICAL REPAIR

Now that José has described some basics of electricity, he will show you how to repair the safety light.

José explains, "A safety light is just like a flashlight, except that it turns on automatically when the room gets dark. A flashlight is a very simple electrical system. All you need to know is how to turn a flashlight on and off to get it to work.

"Here, we have a safety light that does not work. First, I need to understand how the light works inside. To find out why something doesn't work, you need to look at each part. This is called *troubleshooting.*"

Testing

The safety light operates on batteries. José tests the batteries first. To test the batteries, José uses a battery tester. The

Figure 3-4 A battery tester for an individual battery.

tester has two wires. The positive wire is connected to the end of the battery that has a plus sign. The negative wire is connected to the end that has a minus sign.

The battery tester has a needle that shows whether the battery is good or not. The wires, tester, and battery form a **circuit.** The circuit is a path through which electrical current can travel. If the battery works, the circuit is complete and the tester shows that the battery is good. Figure 3-4 shows a battery tester that comes in a new battery package. Some battery testers have two wires that can be hooked up to any size battery.

José sees that the batteries work. Next, he tests the lamp on the safety light. The bulb inside works.

José looks elsewhere in the circuit. Something is causing the light to fail. José says, "When electricity travels through the circuit, it sometimes meets **resistance.**" Resistance is

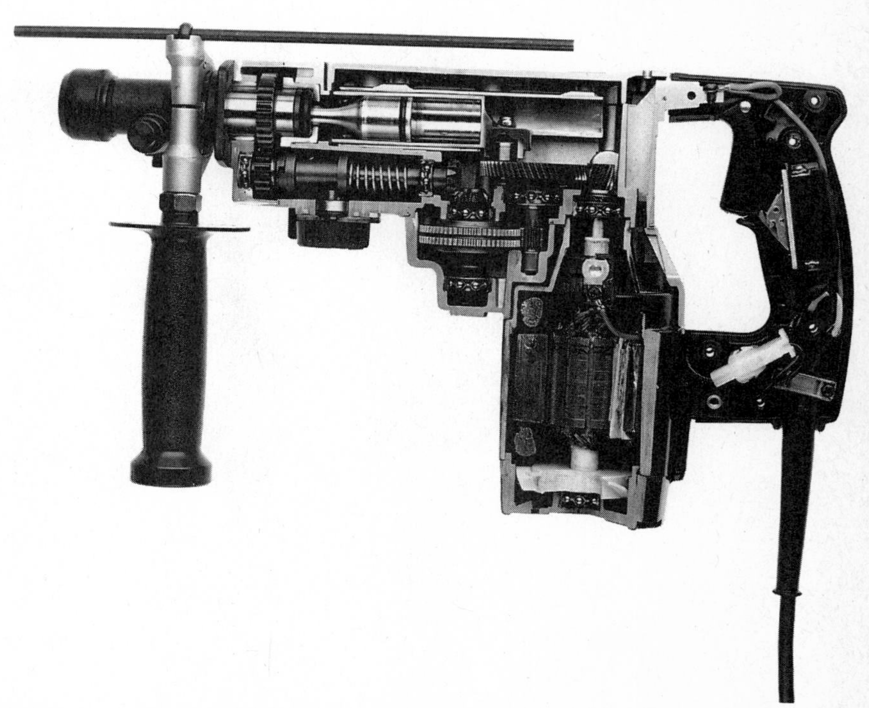

Figure 3-5 A switch inside a hand-held tool.

something that limits the flow of electricity through the circuit. In the case of the light, a switch will cut off or allow the flow of electricity. Most electrical and electronic devices have switches. Figure 3-5 on page 27 shows the switch inside a hand-held tool. The switch is turned on by pressing the trigger in the handle of the tool.

Making a Repair

José looks at the switch. He locates the problem. The wires to the switch have become detached. They will need to be **soldered.**

Soldering is the joining together of two metal parts. It is done by melting the two metal parts so that they join when cooled. A *soldering iron* heats the metals. Today many wire repairs are electronic. Figure 3-6 shows electronic soldering.

Once José solders the wires, the circuit will again be complete. Turning the switch on or off opens and closes the circuit.

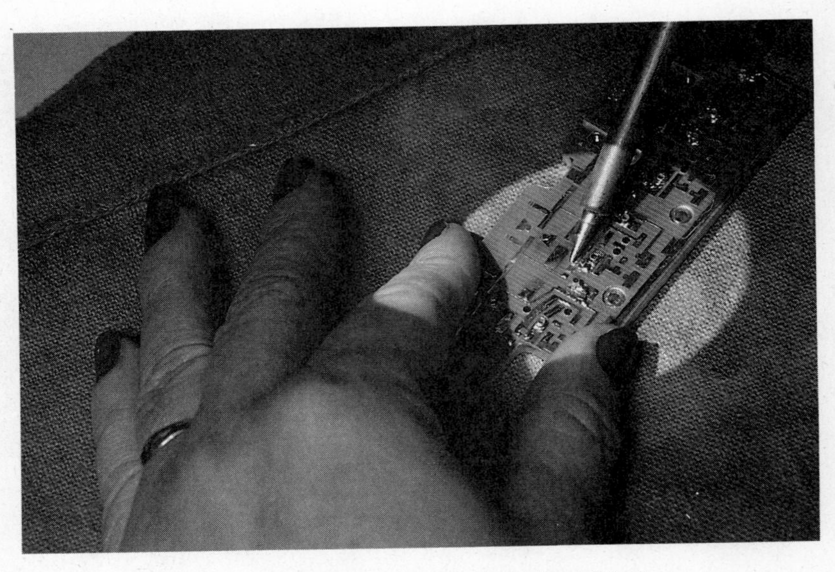

Figure 3-6 Making an electronic connection by soldering.

Using Circuit Diagrams

In the example of the safety light, José made a simple repair. Other items, such as a toaster or waffle iron, get much more complicated. Repair technicians need to refer to diagrams of the inside of an item. These diagrams are called **circuit diagrams.**

Figure 3-7 shows a circuit diagram. Notice the symbols on it. Technicians who work in repair have to read these symbols. Some common ones are labeled in the figure.

The symbols show how the circuit will flow. For instance, if the designer wants to interrupt the circuit at three points, the drawing will call for three switches. These switches can interrupt the flow of electricity by breaking the circuit.

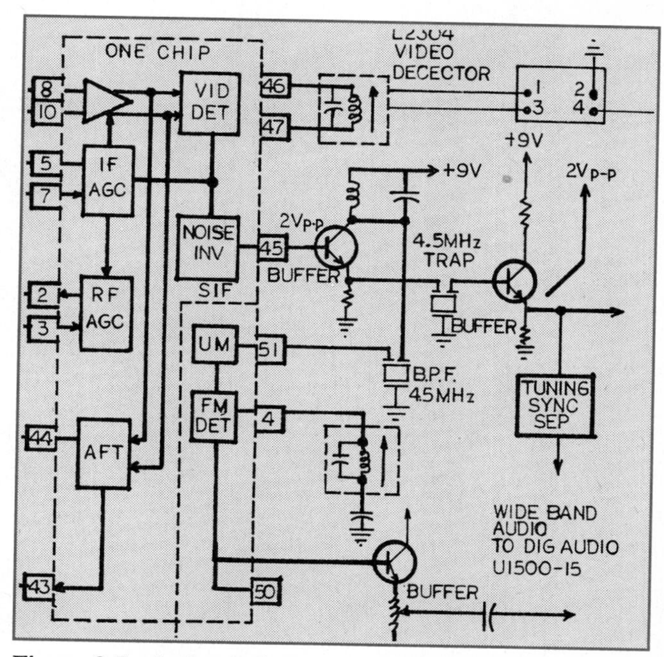

Figure 3-7 A circuit diagram showing symbols.

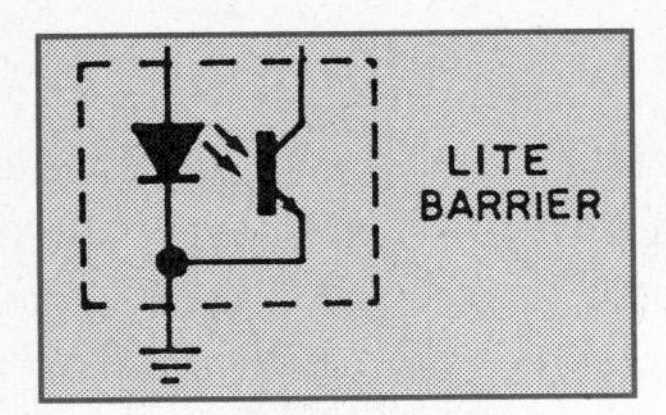

Figure 3-8 The LED symbol.

Figure 3-9 Engineers designing integrated circuits.

Figure 3-10 A designer using a CAD system to design integrated circuits.

Electronic drawings use special symbols. For instance, if you have a clock with a digital display, it probably uses a *light-emitting diode (LED)*. Figure 3-8 shows the symbol for an LED on the circuit drawing shown in Figure 3-7.

Trained technicians learn both electrical and electronic symbols. The symbols make up the special language of electricity and electronics.

Engineers and designers also use the symbols. Figure 3-9 shows engineers designing complicated circuits.

Some computers can set up electronic drawings. Designing by computer is called *computer-aided design (CAD)*. Figure 3-10 shows a designer using a CAD system to design and integrated circuit.

CHAPTER 4

REPAIRING ELECTRONIC ITEMS

Lonnie brings your class next door to the electronic repair department. She introduces you to Susan Watson, an electronic repair technician. Susan talks to you about the work she does. She tells you, "I had to study electronics before I got this job. The most important electronic product today

ELECTRONICS

Electricity is the field of technology that studies the movement of electrons. **Electronics** is the field of technology that deals with electrons moving along **conductors.** Conductors are materials or substances that make an easy **path** for the **flow** of electricity.

Electronic devices also use many **semiconductors.** Semiconductors are materials that fall somewhere between good conductors and good **insulators.** This means that they allow the flow of current but not as freely as conductors do.

Computers

The field of electronics has grown enormously since the development of the **computer.** A computer is any device

Figure 4-1 A computer console on an automobile dashboard. This system gives the driver exact directions to a location.

that can handle **data** or information at high speeds. In addition, the development of tiny **microcircuits** able to handle a lot of data in a small space helped expand the electronics field.

Circuit Boards

Small circuit boards, or **chips**, have allowed the development of **digital** watches, high-speed computers, calculators, and many other machines used in daily life. Some form of computer controls almost all our communications and many of the machines we now use. Many cars now have computer controls for certain systems. Figure 4-1 shows a computer control console in a car. This computer gives the driver directions to a place while he or she is driving.

Understanding Computers

In order to explain computers, Susan uses an everyday example. She says, "Suppose you type the letter *E* on a typewriter with a ribbon. You will get an *E* on the paper. A raised letter will strike an inked ribbon. You will see the inked letter on the paper. Now, suppose you type the letter *E* on a computer keyboard. Something much more complicated happens. The computer gets the information—the letter *E*—as a series of zeros and ones. A capital *E* is 01000101. Each of the zeros and ones is a **bit,** a unit of information. The eight bits in the *E* make up a **byte.** The computer keeps all the bytes of information in its storage, or **memory.** That is, until you tell the computer what to do. For instance, you can see an entire business letter on the screen, called a **monitor.** Then you can correct it. If you need it, you can print it or you can save it for later." Figure 4-2 shows a monitor.

A computer is made up of two main elements—hardware and software.

Figure 4-2 A computer monitor.

Computer Hardware

Computer **hardware** means all the parts of the computer that you can see and touch. It also means all the electronic circuits inside the central processing unit of the computer.

Electroserve does not manufacture computers. But the company does make products that use computers and other electronic parts. One of the most common parts, a **printed circuit board**, is made up of various materials that control the flow of electrons. Engineers design circuit boards to perform certain functions. Figure 4-3 shows a printed circuit board.

Circuit board technology is always advancing. The materials used to make the boards have changed. The component parts are always being improved.

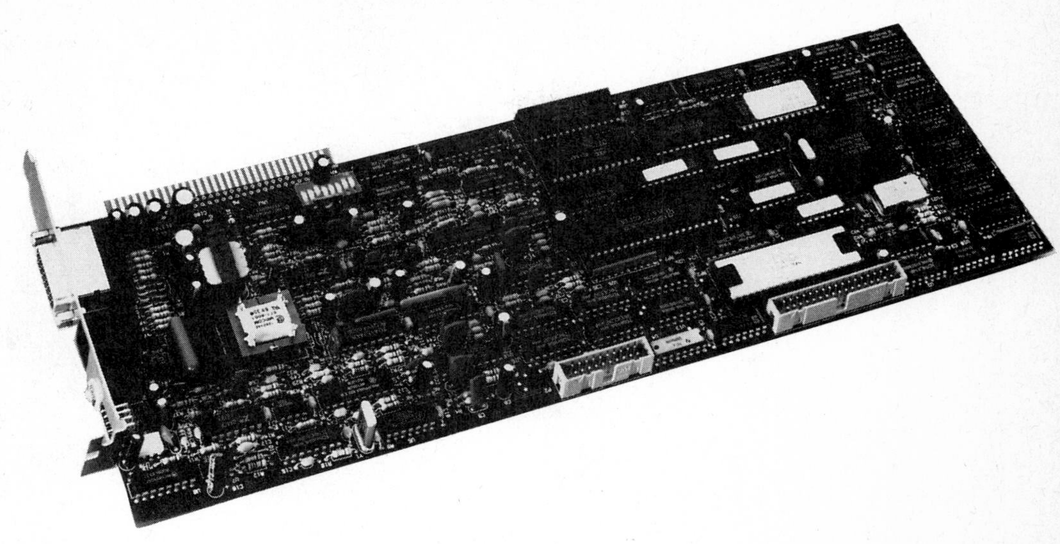

Figure 4-3 A printed circuit board.

Electroserve uses microcircuits in its cash register products. Many of the circuits the company uses are **integrated circuits** (also called **ICs).** This means that on one board, a number of parts are integrated or connected to work together.

Susan tells you that she had to learn about **components,** the parts that make up the whole integrated circuit. Figure 4-4 shows the component parts on a printed circuit board.

Susan says, "When I first went to school, words like *transistor, resistor, diode,* and *capacitor* confused me. I had heard of a transistor radio, but I didn't really know what the words meant."

A **transistor** radio runs with the help of transistors. These electronic components both transfer and resist current. Not too long ago, most radios ran on bulky, unreliable tubes. With the growth in circuit board technology, radios have changed.

A **resistor** resists the flow of current. In order to control the flow and make it go at a certain pace in a certain direction, designers use resistors. Figure 4-5 shows a resistor.

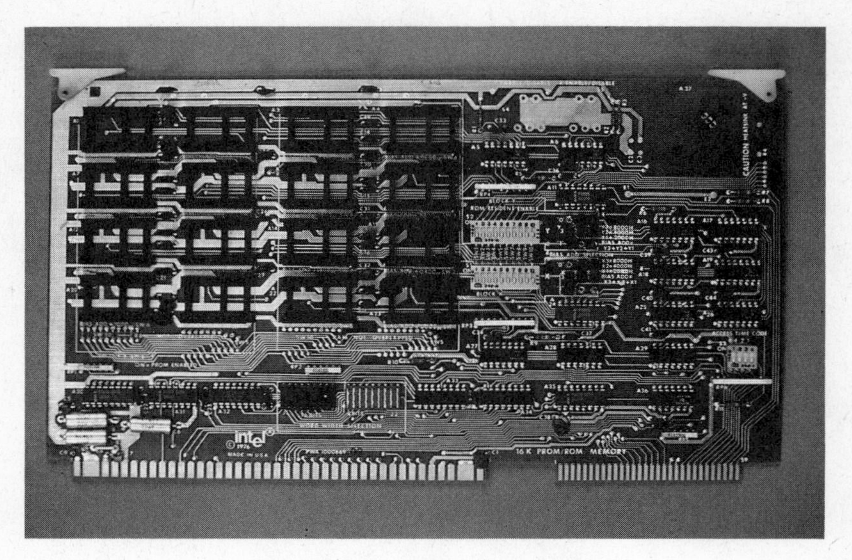

Figure 4-4 Components on a printed circuit board.

A **diode** is an electronic device that allows current to flow in one direction only. Figure 4-6 shows a diode.

A **capacitor** can gather and store electric charges. The charges can be used at a later time. Figure 4-7 on page 40 shows several capacitors.

"Once I learned about all the devices in circuit boards and how they work, I began to understand how electronic equipment uses electrical current to work," says Susan.

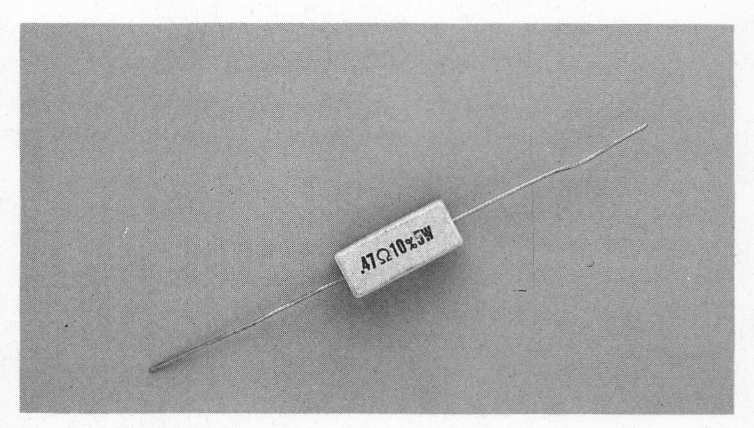

Figure 4-5 A resistor.

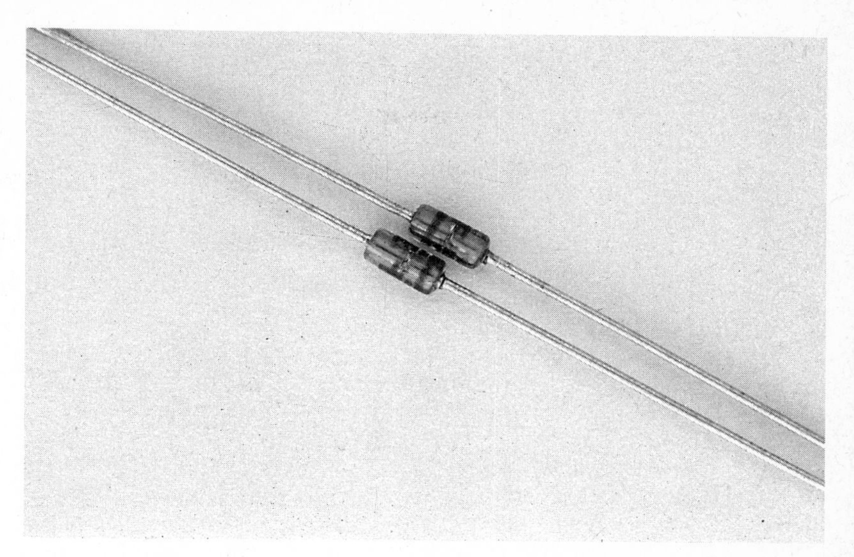

Figure 4-6 A diode.

The cash register pad that Electroserve puts together for Quicky's has a number of circuit boards. When the cashier presses a key for an item on the register, say for example a hot dog, the computer inside tells the register to charge $1.09 plus tax.

Today, printed circuits are getting smaller and smaller. **Microcircuits** are tiny printed circuits that can hold the parts necessary for handling a lot of **data.**

How Small Is Small?

In the last 35 years, the number of devices that can fit on a microcircuit has grown steadily. Millions of circuits can fit on one half-inch circuit. Figure 4-8 shows a microcircuit. Microcircuits are usually placed on **microchips**, or **chips.** Chips are tiny, thin waferlike boards, usually made of **silicon.**

There are two main advantages to having such small circuits.

Figure 4-7 Capacitors.

1. The smaller the circuit board, the smaller the final product can be. In a few years, you may be wearing a television set on your wrist.
2. The second advantage is speed. The shorter the distance electrons have to travel, the faster they can carry current.

Computer Software

Electroserve does make some specialized **programs** for restaurants. The company has developed a computer inven-

Figure 4-8 A microcircuit or microchip.

Figure 4-9 A software display.

tory system hooked into Quicky's cash registers. This keeps track of products, sales, and stock.

Programs that perform tasks are called **software.** In the design department, Electroserve has a software designer at work. The designer upgrades all of Electroserve's programs and comes up with new ones. Figure 4-9 shows software on display in a computer store.

REPAIRING ELECTRONIC MACHINES

Susan is working on a cash register pad. Quicky's sent it in for repair. Their staff said that the machine enters items twice or not at all. Also, Quick's wants to make some changes. A new tax law means that the company has to charge tax differently.

Susan tells you, "Electronic repair often means **upgrading,** or changing for the better. The circuit boards inside the cash register are small. They also have complicated wiring. If I test a machine and find a bad circuit board, I replace it. It costs much less to put in new boards than it does to try and fix them.

"In addition, our computer programmer had to make changes for the new tax law. So after I fix and test the pad, the programmer will make the changes."

Your class leaves the repair area and moves into the manufacturing area. There you will see how products are planned and made.

CHAPTER 5

Working in Manufacturing

The next stop on your class visit to Electroserve is the manufacturing area. You meet Benjamin Orleans, the manager of manufacturing. He shows you a chart of how the manufacturing group operates. Manufacturing has four areas. The first one, planning, has a small group of offices. The second one, the main machining area,

is where parts are actually made. The third and fourth areas are assembly and testing.

To explain the manufacturing process, Benjamin tells you to imagine you work in a cookie factory."Let's pretend the cookie factory makes four kinds of cookies. The first is a plain vanilla cookie, and workers make that from scratch.

The second is a vanilla sandwich cookie. Workers make the cookie part but buy the inside from a candy factory.

The third is a jam-filled cookie. The cookie factory buys the jam to fill its cookies. The fourth is a chocolate-covered cookie that contains a marshmallow. This cookie is assembled from a cookie baked at the factory, a marshmallow from another supplier, and a chocolate glaze from yet another supplier."

Most electrical and electronics manufacturers are similar to the cookie factory. They make some parts. But they also buy parts from different suppliers.

Figure 5-1 shows a cash register that is manufactured by the supplier with some elements bought from other manufacturers. The cash register has both electrical and electronic parts.

Figure 5-1 A cash register being used in a restaurant.

Figure 5-2 A power ratchet with a charger.

In electrical items, there is usually a casing, called a **housing,** a heating element, a timer, and a power supply. The power supply can be inside the item. It can also be a charger outside the item. Figure 5-2 shows a charger that supplies power to the power ratchet.

In addition, there will be parts that perform specific functions. If one factory had to make all the parts, it would require so many machines and people that it would be impractical. That is why most manufacturers do a combination of manufacturing and assembling.

FILLING ORDERS

The manufacturing area fills orders for items. Sometimes, plant managers expect orders and start to manufacture for potential orders. This enables them to have inventory on hand to supply customers quickly.

When orders come in, plant managers look at the numbers. If items are in stock, they will ship them. If they have manufactured the item before, they will tell manufacturing to make the necessary number.

If, however, the item requires a new design or just a change in design, they send the job to design. The designers plan the item. They also get the customer's approval of the final design if necessary. Then design sends the plans to manufacturing's planning department.

PLANNING

The first stage of manufacturing is planning. Planning includes taking designs and instructions and figuring out how the project will go.

The planning department sees if it can make all parts of the item in-house. Often, Electroserve needs to buy parts, make some of the parts, and assemble the item.

Manufacturing keeps its own inventory system. It also does its own purchasing, receiving, and storing of supplies.

Benjamin tells you, "We have to keep track of supplies ourselves. We use thousands of parts. Also, custom-made products require special attention. I use computers to help me handle and do the purchasing for manufacturing's inventory."

Making Purchases

Two basic types of inventory purchases are made by the manufacturing department. When Benjamin buys an item that costs less than $100, he is making a *minor purchase.*

When he buys an item that costs more than $100, he is making a *major purchase.*

Major Purchases

Items that cost more than $100 in the manufacturing department usually are for use in an expensive Electroserve product. Benjamin needs to be sure that such large parts will be guaranteed by their manufacturers. If Electroserve has any problems later with its products because of a part, Electroserve should be able to get some satisfaction from the manufacturer of the faulty part.

Benjamin keeps track of all major purchases in one of the small file cabinets in the office. Records about major purchases may include the following:

- Purchase information (copies of order forms and records of payment)
- Instructions telling how to use the equipment
- Warranties
- Maintenance contracts

Many parts come with a *warranty* stating how long the part is guaranteed to work. For example, a computer part may have a one-year warranty. Suppose a computer part causes a breakdown in one of Electroserve's products before a year of use is over. The part manufacturer will fix the part. Usually there is no charge. Sometimes there is a small fee.

Some of the purchases that Benjamin makes are for the machinery used in manufacturing. He keeps the same kinds of records for that equipment.

Recently, Benjamin purchased a new small computer with several programs. The computer has saved him a lot of time. However, after two months of use, the screen on the monitor began to fade. Benjamin pulled out the warranty. He saw that he was entitled to have the monitor fixed or replaced.

Benjamin called the dealer from whom he purchased the computer. The dealer replaced the monitor. The dealer also

BELROSE INDUSTRIES, INC. **719350**
MAINTENANCE AGREEMENT

Customer: ______________________ Effective Dates __________ to ______

____

____

Contract Equipment	Serial No.	Amount	Product Code

Subtotal ____________
Tax ____________
Total ____________

Customer Acceptance
This agreement, consisting of the terms and conditions stated herein, is hereby approved, accepted, and executed by the respective parties hereto on the date set forth adjacent to their signatures.

X______________________
Customer Signature

____
Title

____
Date

TERMS AND CONDITIONS

1. The terms of this agreement is based upon the anticipate customer usage as stated in the attached specifications.
2. Emergency service calls will be performed at no extra charge providing such calls are made during normal business hours. Overtime charges at Belrose's then current rate will be charged for all calls made outside normal business hours.
3. This agreement covers all routine, remedial, and preventive maintenance service.
4. With the exception of consumable parts, all parts are included under this agreement.
5. This agreement is not assignable or transferable.
6. This agreement will not apply to any equipment lost or damaged through accident, abuse, misuse, theft, neglect, acts of third parties, fire, water, casualty or any other natural force and any loss of damage occurring from any of the foregoing is specifically excluded from this agreement.

Figure 5-3 A maintenance contract.

told Benjamin that he was leaving the monitor on for too many hours during the day. This can hurt a monitor. So now, when Benjamin is not using the computer, he shuts off the monitor button.

A *maintenance contract* is a written agreement between the buyer of a piece of equipment and another company. The buyer pays a certain amount of money to the seller, who agrees to keep the piece of equipment in working order. The manufacturing department has maintenance controls for all its major machines. The contract guarantees fast service at a fixed cost. Figure 5-3 shows a maintenance contract.

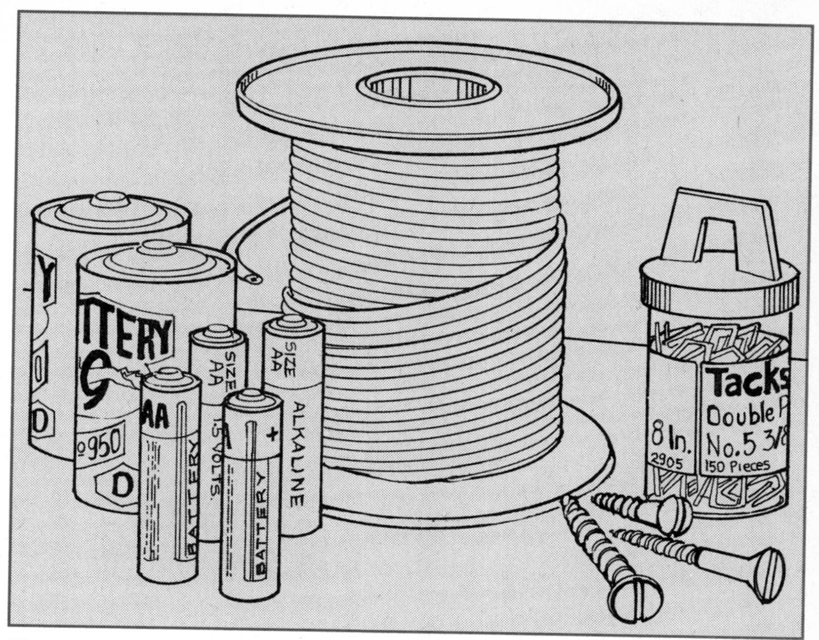

Figure 5-4 Minor purchases.

Maintenance contracts are expensive. Benjamin files the repair information for each year in the equipment file. He enters the date and cost of each repair in a log book. Every year the managers reevaluate the cost of a maintenance contract. For a few pieces of equipment, Electroserve has chosen not to buy a maintenance contract, but such contracts have been worth it for most of Electroserve's essential equipment.

Minor Purchases

Minor purchases cost less than major purchases. They are used up quickly. Most of manufacturing's minor purchases are small items, such as microcircuits, batteries, screws, and wiring. Figure 5-4 shows some of the minor purchases that Benjamin usually makes.

In the office area, the receptionist keeps track of office supplies, such as envelopes, printed forms, notebooks, and pens. For those kinds of supplies, an exact *inventory system* is not needed. If the department ran out of pens, they could purchase them the same day. They would not have to stop working.

In the manufacturing department, Benjamin has a very exact inventory system. You cannot just go and buy printed circuit boards with exact specifications. They have to be ordered. There are many items, such as housings, that have to be made by other manufacturers before Benjamin can receive them.

Some of the items used by manufacturing can be bought at a local hardware store. It is not necessary to keep track of the exact number of screws-in house. However, Benjamin must keep track of a lot of items in his inventory so that he knows how many have been used.

Inventory Control

Keeping the right amount of supplies or parts on hand is known as *inventory control*. Managers need to know when to order inventory and how much to order.

When to Order

How does Benjamin know when to order manufacturing supplies? He does not want to order an item when he has plenty on hand. That wastes storage space. It also uses company money too far in advance of when an item is needed. In most companies, cash flow, or the availability of money, is usually a big concern. All the managers have been told to keep their budgets as small as possible.

But, if Benjamin waits too long, he may run out of an item. That would waste expensive worker manufacturing time. One of the main tasks of a person in charge of inventory control is to decide when it is time to order an item.

To decide when to order an item, you need certain facts. You need to know how fast your office uses up the item you are ordering. You also need to know how long it usually takes the company to deliver the items. Suppose you know that it takes three to four weeks to order plastic for menu boards. You will order the plastic at least four weeks before you run out.

How Much to Order

Benjamin needs to take three things into account when deciding how much of each item to order.

1. The amount of storage space available in the company.
2. How long an item can stay on the shelf before the company can no longer use it. This is known as *shelf life.*
3. Whether the company will save money by ordering large amounts.

Some manufacturing supplies have a shelf life. For instance, batteries do not last more than a year or two. If the manufacturing department rarely uses AAA batteries, it would be wasteful to order a large quantity of them. On the other hand, if C batteries are used in all safety lights and several other products, then a large quantity of them should be kept on hand. The shelf life of the C batteries does not matter because they are used so quickly. The shelf life of the AAA batteries does matter because they are used so rarely.

Another type of shelf life is the usability of an item. Certain components lose their effectiveness because they are replaced by more advanced products. This is especially true in the field of electronics. Products keep getting smaller and more efficient, and often even less expensive.

For example, let's say that Electroserve has been using one type of printed circuit. A salesperson presents the company with a new, more efficient, even cheaper type of printed circuit. Electroserve would probably want to change over to the new one as quickly as possible. If the company had a huge stock of old circuit boards, it would not be able to change over very quickly.

So for electronic items, especially, Benjamin orders only the circuit boards that manufacturing needs for about three months. This way, Electroserve's designers can be up-to-date with their product.

Many companies offer discounts when you buy large quantities of an item. For example, a company that makes batteries, gives the wholesale cost of a C battery as 50¢. The price of 50 batteries is $23. The price of 200 batteries is $90. Benjamin buys 200 at a time to take advantage of the discount.

Keeping Track of Supplies

There are many different ways to keep track of supplies. The manufacturing department uses a system of *inventory cards.* It also uses an *inventory report form.*

Inventory Cards

Benjamin makes up an inventory card for each item that he orders. Figure 5-5 shows an inventory card. Inventory cards

ORDER (item name): Circuit Board M-18H

Order Quantity: 50 Reorder Point: 50

Date	**Quantity**	**Date Received**

Order Source: J.B. Electronics

Figure 5-5 An inventory card.

name the item. They show the quantity to order and the *supplier,* the business that makes the item.

Some of the inventory cards shows a *reorder point* for that item. Look at the inventory card for the F-23X printed circuit board. The reorder point on the card is 100. When there are only 100 of these circuit boards left, Benjamin will order more.

If you were to look at the inventory card for 2-inch nuts and bolts, you would not see a reorder point. No one counts the nuts and bolts. They are purchased in huge quantities. Benjamin can tell by looking at the container when the supply is getting low.

The Inventory Report

How will Benjamin know when there are only 100 F-F-23X printed circuit boards left? Sometimes he finds out when an assembly clerk notices that the storage room is running low on an item. If necessary, Benjamin will order that item right away.

Another way Benjamin finds out when stock is low is the inventory report form. Every month, Benjamin gives an inventory report form to the assembly clerk. The clerk counts the stock on hand for the supplies listed on the form. She fills in the form. Using the report and the inventory cards, Benjamin decides which supplies to order. Figure 5-6 shows an inventory report form.

Ordering Supplies

To order supplies, Benjamin fills out a purchase order. At the top of the purchase order, he writes the name and address of the supplier. He writes the date of the order. The purchase order has columns in which to fill in the number of items being ordered, a description of each item, and the unit price. *Unit price* means the price per item. Benjamin also fills in the total dollar amount of the order.

INVENTORY REPORT		
DATE: June 1, 199X ITEM	STOCK ON HAND	TO ORDER
Printed Circuit Boards		
F-23X	200	______
K-150	40	______
M-17G	20	______
R-12	95	______
M-18H	20	______
Microcircuits		
P-11	720	______
XL33	90	______
ML27	50	______
R449B	1000	______
A-17	75	______

Figure 5-6 An inventory report.

On orders over $250, Benjamin needs to have the purchase order signed by a plant manager, so that the office has some idea of how much is being spent. Plant manager approval also controls excess ordering.

Each purchase order has a different number. The original form is sent to the supplier. A copy is kept on file at Electroserve. If a supplier has a question about an order, the supplier will refer to the number of the purchase order. The copy is used to check the order when it arrives. Figure 5-7 shows a completed purchase order.

Supplier: SuperElectric

Ship To:
Electroserve, Inc.
1306 Industrial Highway
San Rafael, CA 90005
(415)516-4000

Address:
Corporate Drive
Eden Prairie, MN 55578

Date: June 2, 199X

Quantity	Description of item	Unit Price	Extension Price
5	R12 minibulb	$7.50	$37.50
100	Y23 circuit board	$12.95	$1295.00
150	M1128 wire	$2.50	$375.00
25	Q111 transformers	$20.50	$512.50
50	X34 capacitors	$1.25	$62.50
100	RX45 switches	$2.95	$295.00
		Subtotal	$2577.50
		Shipping	$32.50
		Total	$2610.00

Figure 5-7 A purchase order.

Receiving Supplies

When supplies come in, Benjamin signs for them and calls the assembly clerk. She opens the boxes. Each order comes with an invoice or a packing slip. An invoice lists the goods inside the box and the price of each. The clerk compares the invoice with the purchase order for that shipment. She makes sure that the supplier has shipped the right items. The invoice is also a bill. After checking the invoice, The clerk gives it back to Benjamin. Figure 5-8 shows an invoice.

Invoice

DATE: June 1, 199X

SOLD TO: Quicky's #14
1212 North Alameda
San Rafael, CA 90005

PARTS:

Quantity	Item	Unit Price	Extension Price
1	Menu Board (small for pick-up counter)		
2	Speakers		
1	Register pad with new tax updates		
2	Safety lights		
		Total Parts	
		Tax	
		Total Charge	

Figure 5-8 An invoice.

A packing slip only lists the contents of the package. The supplier sends a bill separately.

Sometimes, an order cannot be filled because the item is "on back order." This means that the supplier has none in stock. The supplier is waiting for its own manufacturing department to make more of the item. When this happens, the assembly clerk makes a note on the inventory card. If she can't get the item from the supplier in a reasonable amount of time, she may look for someone else who can fill the order. Figure 5-9 shows a clerk checking supplies.

Storing Supplies

Suppose four shipments come into manufacturing. One box contains 50 housings. Another box contains 200 batteries. The third box contains 400 printed circuit boards. The last box contains wiring.

Each item has an assigned place on a shelf in the assembly storage area. The boxes are moved to the area. The clerks

Figure 5-9 Checking supplies as they come in.

put the stock in its proper place on the shelves. They store new supplies behind or under old supplies so that the older ones will be used up first. This is called *rotating the stock.*

A good system of inventory control is a must for any company. Benjamin is very proud of the system he set up in the manufacturing department at Electroserve.

Making the Product

Once the manufacturing area figures out which parts can be made in the plant, the staff goes ahead. Managers assign the jobs to the various workers at the machines. They give them the planning drawings, the materials, and the exact specifications. Usually, a sample gets made and approved.

The planners estimate how long it will take to make each part. They also find out how long it will take for the ordered parts to come in. After they have estimated the time, they plan for the assembly of the products. In the next chapter, you will meet an assembly technician.

The planners can then tell the plant managers when items can ship. The managers can notify the customers.

The planners assign the product to the workers in the manufacturing department. Most workers on machines have experience. Manufacturing machinery can be dangerous. Also, experienced technicians are needed to make sure that the product will be exactly as planned.

Using Machinery

Manufacturing machines can perform many different functions. Some mold materials, such as plastic. Others stamp patterns or names onto a material. Still others shape and cut. In our cookie factory example, there are machines to shape, bake, fill, coat, and package cookies. At Electroserve, the machines mold, cut, shape, paint, and stamp materials. Figure 5-10 shows some common manufacturing machines.

Figure 5-10 Manufacturing machinery.

Quality Control

As technicians make parts, others check the size and quality. *Quality control*, or the making sure of quality, is an important job. A company can lose customers fast if it has poor quality control. On the other hand, a company with a good reputation will get more customers. A quality-control technician usually has some training in the science of materials.

Assembly and testing are two separate departments within manufacturing. Your class will meet some workers in both of these areas.

Figure 5-11 Computer monitors being packed in styrofoam before shipping.

SENDING OUT THE PRODUCT

After all four stages of manufacturing are complete, the item goes to shipping. The shipping department packages each product. Electrical and electronic packaging have become very important. Products are getting smaller.

Each circuit board can control many functions. The packaging protects these delicate parts. Figure 5-11 shows computer monitors being put into packaging made of Styrofoam, a dense, protective material.

The shipping department takes care of sending out the products. Shapping clerks also keep track of all items sent. They send that information back to inventory and billing. The inventory department keeps track of items on hand.

Invoice

DATE: June 1, 199X

SOLD TO: Quicky's #14
1212 North Alameda
San Rafael, CA 90005

PARTS:

Quantity	Item	Unit Price	Extension Price
1	Menu Board (small for pick-up counter)	$1375.00	$1375.00
2	Speakers	$48.50	$97.00
1	Register pad with new tax updates	$962.75	$962.75
2	Safety lights	$72.50	$145.00
		Total Parts	$2579.75
		Tax	$129.46
		Total Charge	$2709.21

Figure 5-12 An order form and bill.

❑ GETTING PAID

The billing department makes sure that customers get accurate bills. Billing clerks also check to see that the shipped items match the original order. Figure 5-12 shows an order form with pricing on it. The billing department sends this form to the customer. The customer has 60 days to pay the bill.

The billing department is part of the accounting area. The accounting workers keep track of outstanding bills, called *receivables*. They follow up with customers to make sure that Electroserve gets paid on time.

CHAPTER 6

ASSEMBLING ELECTRONIC PRODUCTS

Benjamin brings you into the assembly area. There you meet Stephanie Jones, an assembly technician. Stephanie will show you the department. She will also show you how she assembles a product.

"There are several stages in any assembly operation," Stephanie tells you, "You've

seen how the inventory department keeps track of and orders parts. Now you'll see for yourself how the workers put the parts together."

First, Stephanie shows you the shelves in the supply room. There are hundreds of parts all labeled with numbers and a bar code. One of Stephanie's jobs is to make sure that the inventory department knows how much stock it will need. Figure 6-1 shows a supply area in a large manufacturing company.

Getting the Parts Together

In the assembly supply area, you meet two clerks. They are putting together **kits.** A kit contains all the parts needed to assemble a product or a part of a product. For instance, a safety light may need only one kit. A cash register pad will need several kits. Figure 6-2 shows a technician assembling computer monitors. Notice the kits next to him.

Figure 6-1 A supply area in a large manufacturing area.

Figure 6-2 A technician assembling computer monitors.

Some kits consists of many tiny parts. Other kits have large parts and small parts. For instance, a computer housing may measure 2 feet by 3 feet. The cards, or circuit boards, that run it may be slightly thicker than several sheets of paper and maybe 6 inches long by 3 inches wide.

The clerks get all the kits ready for each assembler. If assemblers had to find each part, they would waste too much time. The clerks must handle parts carefully. They also have to put the right parts in the right kit. If not, the assembler will

Figure 6-3 Assembly technicians at work in a large manufacturing company.

waste a lot of time. Figure 6-3 shows assembly technicians at work in a large company. All the technicians have the needed parts at their stations.

Job Numbers

The assembly clerks know exactly what parts go into each kit. They have a list of *job numbers*. Each job number has a master list of parts. Some customers order items with variations or additions. The order form tells the assembly clerk what parts to include or take out.

In Electroserve's system, each clerk checks off items on a list as each kit is gathered. This system and the use of job numbers help keep the work accurate.

Assembling the Parts

What do you think of when you hear the term assembly worker? You may have a picture in your mind of an assembly line in a factory. Suppose the factory makes dolls. The workers are lined up next to a moving belt. On the belt are the dolls. As the dolls move along on the belt, each worker attaches a single part. One worker may attach a leg, another worker may attach an arm, and so on.

Electroserve's operation is not at all like that. Each assembly technician works on a complete product. Because of the complicated nature of some of Electroserve's products, the company prefers to have each trained technician complete whole products. This helps in quality control.

"We line up the devices we are assembling," Stephanie tells you. "But, we don't use an assembly line to put them together."

As you go into the assembly area, you see long tables. On each table are six large plastic boxes. Three assembly workers holding screwdrivers are standing in front of three of the boxes.

The boxes are cases, or **housings,** that hold the parts of a machine. Housings are as large as needed to hold all the parts. Housings for wristwatches are very small. Housings for television sets are as large as the television. Figure 6-4 shows the housing for a television set.

Figure 6-4 Housing for a television.

The housing has holes and slots. Screws go through the holes to hold parts of the interior. Slots can hold parts such as computer boards or cards. Each product has a different inside. The assembly worker has a drawing that shows how the product should look.

An assembler needs to follow instructions exactly. Each piece must go in exactly as noted on the instructions. A designer has already figured out the makeup of the item. Electrical and electronic items require exact setups so that the current flows as designed.

Assemblers often have to wire the inside of a product. The connections made by wires have to transmit current for the life of the product. A loose wire will cause problems.

Basic Wiring

Loose wires remind Stephanie to discuss wiring. She says, "You are probably familiar with electrical wiring. Your television has a wire that plugs into an outlet in the wall. It also has some internal wires that **conduct** electricity through the various parts of the television." Figure 6-5 shows a wire coming out of a drill. It is plugged into an electrical outlet to get the electricity needed to operate the drill.

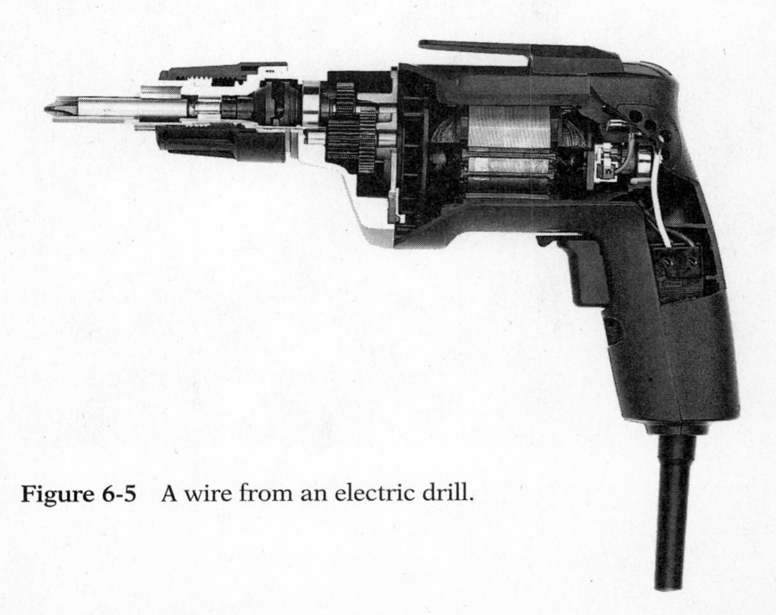

Figure 6-5 A wire from an electric drill.

Stephanie continues, "What you may not be as familiar with is electronic wiring. Printed circuit boards have 'printed' wires. That means that thin strips of a **conductor**, such as copper or silver, are put on the board. These materials allow the electrons to flow in the path laid out on the circuit board." Figure 6-6 shows wiring on a printed circuit board.

Stephanie says that in school she had to learn about the types of electrical and electronic wire. She also learned about the different materials in wire.

Electrical Wire

Solid wire does not bend easily. It connects some machines to outlets. It also completes the circuits inside of walls. Once solid wire is put in place, it usually stays there. Solid wire cannot be moved very often. If bent or flexed often, it will break.

Figure 6-6 Printed wire on a circuit board.

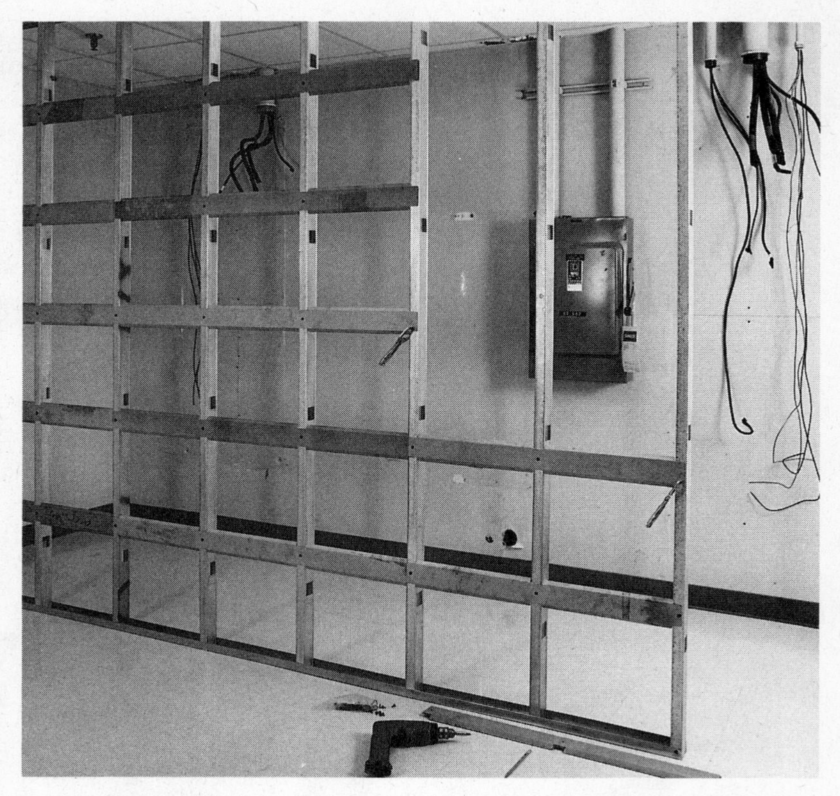

Figure 6-7 Stranded wire at a construction site. (Courtesy Armstrong World Industries, Inc.)

Wire that must be flexed is made up of **strands.**The strands are held together by a covering. Figure 6-7 shows stranded wire at a construction site. Usually the number of strands depends on the use of the wire.

Groups of stranded wire held together are called **cable.** Cables hold such things as telephone wires together. Figure 6-8 shows a cable.

For safety, wiring outside a device is usually insulated with a covering. Some wires have a varnish coating. Some have a covering that does not conduct electricity. Some have a braided covering. Wires inside devices may be left bare.

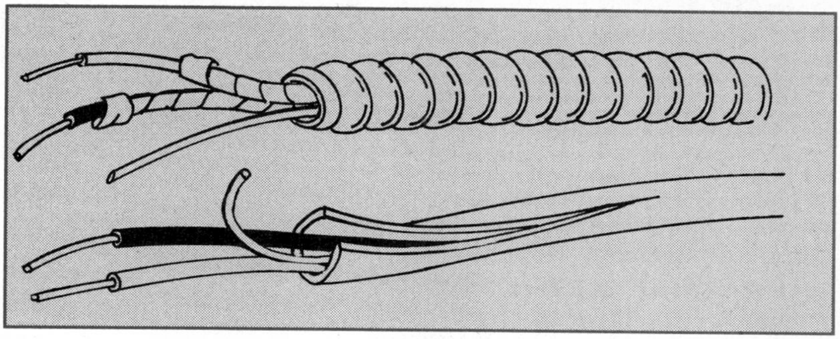

Figure 6-8 Cable showing stranded wire held together.

Manufacturers can choose from hundreds of types of wires. They also have a choice of materials. Different conductors have different **resistance.** That means that some wires allow the electricity to flow more easily than others.

Actually, four things determine resistance. The first is the size of the conducting wire. A larger size allows more electrons to flow. Imagine a six-lane highway as opposed to a two-lane road. The six lanes allow much more traffic to pass through at one time. A large wire allows many more electrons to pass through at one time.

The length of the wire affects resistance. The longer the wire, the more the resistance. The electrons have to travel a longer distance between two points.

Certain materials have greater resistance than others. Steel has about ten times the resistance of silver. Scientists are always developing new conductors. A new group of materials, called **superconductors,** have virtually no resistance. They will play a large role in the future of electricity and electronics.

The temperature of the material also affects resistance. The hotter the material, the greater the amount of resistance. Carbon is a material that can reach very high temperatures. It has a very high resistance. Manufacturers of automobiles use carbon in certain parts. This allows the wire to reach very high temperatures.

Manufacturers and users of wire may have other special needs. A company that uses stranded wire, or cable, is the telephone company. Telephone installers constantly connect wires. Usually telephone wire is made up of coated colored strands. The installers know that certain colors stand for certain connections. This color coding helps them avoid costly and dangerous mistakes.

Connecting Wire

In order for wire to conduct electricity, a metal-to-metal connection must be made. A wire can be wound around another metal, such as a screw. It can be **soldered** in place. Wires taped together can conduct electricity.

Figure 6-9 shows a technician working on power lines. He is making sure that all the wires are connected correctly. The connected wires allow the electricity to flow through the lines.

Figure 6-9 A technician working on power lines.

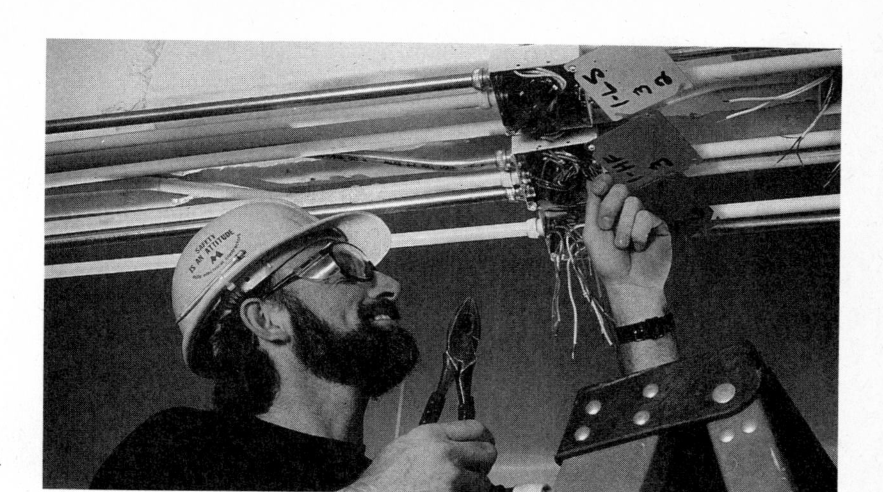

Figure 6-10 An electrician taping wires together for a ceiling fixture.

Figure 6-10 shows an electrician taping wires together for a ceiling fixture. The electrician has wound the wires together. The tape prevents anyone working on the fixture from getting a shock.

Since electricity has many dangers, national standards have developed for its use. The *National Electrical Code* has wiring rules and regulations. Government inspectors make sure that wiring in the areas of manufacturing and building meets the code's standards.

Products that meet national standards have a symbol on them. One common electrical symbol is *UL,* which stands for *Underwriter's Laboratories.* This shows that the electrical parts meet a certain standard. Figure 6-11 shows a *UL* symbol on an electric device.

Figure 6-11 The UL symbol.

Electronic Wire

Electronic wire usually consists of copper or silver "printed" on a circuit board. Printed circuit boards vary greatly. For some complicated machines, the circuit boards actually have several layers with metal patterns painted between the layers.

In manufacturing and in repairing printed circuit boards, a soldering process is used. Since the boards are so small and complex, the process is a precision one. Figure 6-12 shows precision soldering of electronic parts.

Figure 6-12 Precision soldering of electronic parts.

Figure 6-13 A fiberglass printed circuit board.

The board itself must not conduct electrons. Fiberglass is used for boards because it is a good **insulator.** Any current that touches it will not pass through it. Figure 6-13 shows a fiberglass printed circuit board.

The electronic parts, called **components,** are attached to the board. The "painted" wire connects the components. The designers and engineers originally figured out the placement of all wires and components. As the current passes through the components, switches are activated. The switches tell that part of the machine how to operate. Figure 6-14 shows switches on the components of a printed circuit board.

Figure 6-14 Switches on components on a printed circuit board.

Inserting Printed Circuit Boards

Technicians who handle printed circuit boards must know how to do it carefully. Any particles, such as dust, could disturb the delicate switches. Technicians usually work in a clean, dust-free environment. They often wear gloves while handling circuit boards.

Many machines and devices use printed circuit boards. The designer plans for the location of the boards. The housing has holders, or slots, for the boards. The boards are gently fitted into the slot. A tiny screw sometimes holds the side of the board in place.

FINISHING THE PRODUCT

Once the assembler inserts all the parts, the product will go to the testing department. Before the assembler closes the housing, he or she will test the power supply. The assembler will plug the unit in to make sure that the power works. Then the unit can go to testing to see that it functions exactly as needed.

CHAPTER 7

TESTING ELECTRICAL AND ELECTRONIC PRODUCTS

Next, Benjamin takes you to the testing area. He explains, "We test all products before they leave the plant. This avoids customer problems. It also helps us control the quality of work produced in the plant."

Your class meets David Valenti, a testing technician. He explains all the equipment you see in the testing area. He also says, "You would think that we wouldn't have to test every product. Since we do so many products that are exactly alike, we should be able to test one or two. But that is not how it works. Once you have experience, you know that electrical and electronic products are very sensitive. One tiny mistake and everything falls apart."

Electroserve has a fine reputation. Its customers expect a good-quality product. They also expect fast, efficient service if they do have a problem.

The testing department is the place where problems can be avoided. David explains that each product has a testing procedure. "Over the years, we keep track of customer complaints. We keep refining our testing procedure to fix problems that crop up over the life of the product.

"Our testing procedure also reflects the concerns of the design and engineering department. The staff have specific ideas about how a product should perform. The testing department makes sure that the products meet those standards." Figure 7-1 shows a testing department.

Figure 7-1 A testing department.

Testing Procedures

For each product, the testing department sets up a system. Electroserve's products must have their functions tested. For instance, a technician tests Electroserve's safety light by putting it in a dark space to see if the automatic lighting system works. Another technician checks Electroserve's cash register pad by putting it on a machine and seeing if the numbers come up on the screen correctly.

Each product has its own testing checklist. In this way, Electroserve makes sure that each technician performs all the necessary tests. Figure 7-2 shows a testing checklist.

Electroserve Testing Department Checklist

Item: Register Pad for ZXS2 Cash Register (Quicky's standard) Technician: ____________

Serial no: ____________ Date: ____________

Standard tests

1. On/off switch operational ____________
2. LED display can be seen from both sides ____________
3. Power flows to all buttons. Press each one separately. Responses from every button ____________
4. All PCB's are held tightly within casing ____________
5. Use attached sheet to input specific items. Output must appear exactly as shown ____________
6. Final total is accompanied by two short beeps ____________
7. Leave display on for 2 minutes. Light-saver display fades ________
8. Hit enter to restart display ____________
9. All parts cleaned before shipping ____________

Special tests

(Note: All special test questions should be referred to the designer.)

I guarantee that all of the above tests have been performed satisfactorily.

____

____

______________________ Technician ______________________ Supervisor

Figure 7-2 A testing checklist

Each technician checks off each item on the checklist. The supervisor then reviews the checklist and signs it. This system tends to make everyone more responsible.

Each electronic product has a serial number. If there are problems later on, the managers come back to the testing department to find out how something slipped through. The serial number is on the top of each checklist.

Testing technicians must examine all electrical and electronic equipment. Most of this testing requires the use of **testers,** or **testing devices.** These devices measure the flow of current and resistance to that flow. The technician determines if the current and resistance amounts are correct.

Testers

David shows you the testing area. You see lots of machines with numbers on them. David explains, "In school, I learned how to use many measuring devices. Here, we perform some simple function tests. If we find a problem, we measure the current and how it flows to find out where the problem lies."

Figure 7-3 Electric meters on a two-family house.

Meters

A meter measures something. Parking meters measure time. Electrical meters on a house measure the amount of electricity used by the residents. Figure 7-3 shows electrical meters on a two-family house. Each part of the house has a separate meter so that each family pays for its own electricity.

Meters show numbers on a dial. The display can be either **analog** or **digital.** An analog display shows the numbers with marks between them to indicate a certain amount. A digital display shows only one amount, or measure. A good way to understand the difference between the two displays is to compare two common types of watches. An analog watch has a circular dial with marks or numbers all around. Hands point to the correct time on the dial. A digital watch displays the time in one set of numbers or **digits.** Figure 7-4 shows both types of watches.

Some meters have specialized uses. A thermometer measures temperature. *Thermo-* means "heat." Similarly, there are specialized electrical and electronic meters that have different uses.

Figure 7-4 An analog watch (left) and a digital watch (right).

Ammeter

An **ammeter** measures the **amps,** or **amperes**. An ampere is one unit of measure of current flow. Figure 7-5 shows an ammeter.

Current flows in one of two ways—direct current or alternating current. **Direct current (DC)** flows in one direction only. An example is automotive circuits. The current flows directly from the battery to the circuits.

Alternating current (AC) regularly changes its direction of flow. Most electrical current to buildings and houses in the United States is alternating current.

Figure 7-5 An ammeter.

DC ammeters measure direct current flow. For instance, you can check the current at various points in an automotive circuit using an ammeter.

AC ammeters check alternating current flow. For instance, you can check the current flow in a television using an ammeter.

A trained technician knows how to hook up an ammeter. The technician must make sure that the circuit is open at the point being tested.

Voltmeter

A **voltmeter** measures the **voltage.** Voltage is the force that causes electrons to flow. A battery may have a label indicating 9 **volts.** That means that it has a voltage that equal 9 units of force. A 9-volt battery can light a bigger flashlight than a 1.5-volt one. Figure 7-6 shows five commonly used batteries of different voltages.

Figure 7-6 Five common battery sizes.

Figure 7-7 A voltmeter.

The technician learns how to use a voltmeter. The voltmeter must be connected properly. Usually the wires from the voltmeter are colored black for negative and red for positive. Figure 7-7 shows a voltmeter.

Ohmmeter

An **ohmmeter** measures **ohms.** An ohm is a measure of resistance. The resistance in a circuit prevents the flow of electrons. Figure 7-8 shows an ohmmeter.

Multimeter

A **multimeter** combines the functions of an ammeter, a voltmeter, and an ohmmeter. Some multimeters measure

Figure 7-8 An ohmmeter.

both AC and DC currents. Since these testing devices can do so much, they have become very popular. Figure 7-9 shows two types of multimeters.

Other Testers

Technicians like David learn to use many other testing devices. Some testers are very specialized and can save technicians time. For example, with a testing device, a technician can locate something in a minute without having to rip a product apart.

Oscilloscope

An **oscilloscope** allows the technician to see the wave patterns of an electrical current. This visual examination of how

Figure 7-9 Analog (left) and digital (right) multimeters.

current flows can help in testing. An oscilloscope is a very complicated and useful instrument. Figure 7-10 shows an oscilloscope.

Battery Tester

Some testing devices check one part, such as a battery. When the technician hooks a battery up to a tester, the reading will immediately show if the battery can still operate. Often this simple test can save other time-consuming tests. Figure 7-11 shows the reading on a battery tester.

Each industry has testing devices for its electrical and electronic products and equipment. Sometimes, these testers can be a combination of several devices. Other times, the company has designed its own device in response to a particular testing need.

Wherever electrical or electronic technicians eventually work, they are likely to use testing devices.

Figure 7-10 An oscilloscope.

❑ FUNCTION TESTING

Electroserve performs many function tests. Function tests show that a product can do all its intended tasks. In this type of testing, the technician pretends to be a customer. He or she uses the product in ways that a customer might.

An example of function testing is to see if a cash register enters the products and prices correctly. For instance, if the technician presses a button marked *single hamburger*, then the price $.95 should appear on the display along with the initials SH.

The technicians do not guess at what the functions might be, though in some cases, the functions are obvious. At Electroserve, the designer includes a function checklist along with the specifications for the product. The checklist has a space for special comments. This is used when customers order a custom-made product. Figure 7-12 shows a function checklist. The serial number is put on the top. The techni-

Figure 7-11 A reading on a battery tester.

cian and the supervisor must both sign the list in order to approve the product.

Some machines can be hooked up to a computer for function testing. For instance, an automotive emissions tester will hook up a computer to the exhaust pipe. The computer tells the tester how the car's system is performing. Figure 7-13 shows an emissions test being done on a car. One technician is hooking up the reading device to the car's exhaust. The other technician is reading the data on a computer.

FUNCTION CHECKLIST

Electroserve Function Checklist

Item: Cash Register Pad for Quicky's Drive-thru service   Technician:____________

Serial No:__________   Date:____________

Function tests:

1. Shows large display for order service for each item pressed____________
2. Beeps once at the end of every item ordered____________
3. Beeps twice at the end of an order (when enter is pressed)____________
4. Enters tax after enter key is pressed____________
5. Overrides tax and final total is ESC/ADD is pressed____________
6. LED display fades after two minutes____________

Special function tests
(Refer special function test questions to the designer.)

____

____

____

I guarantee that all of the above tests have been performed satisfactorily.

________________________   ________________________
Technician   Supervisor

Figure 7-12 A function checklist.

Some testing areas now look like computer labs. Each testing technician sits at a computer, hooks up products, and reads the results. Figure 7-14 on page 91 shows a computerized testing station.

FINAL CHECK

The quality-control area has checked all the parts. The testing department has checked all the functions and tested

Figure 7-13 A computer testing station at a state-run automotive emissions testing center.

Figure 7-14 A computerized testing center at a manufacturer.

the current. Now, it is time to ship the product. The manager of the testing department looks at the products for a final check.

Before products ship, someone in charge should look at them. Do they have all the proper labeling? Does the surface finish look brand-new? If you were the customer, is this how you would like to receive the product?

As with every step along the way, the workers contribute to the company's reputation. If you send out a fine product that works, customers generally bring their return business to you.

CHAPTER 8

ON-SITE SERVICE

Your class is going to the Quicky's Restaurant on the next block. Joyce Almenta goes with you. She is from Electroserve's service department. She will meet an electrician there and assist him in installing new menu boards. Before leaving Electroserve, Joyce has made an appointment with

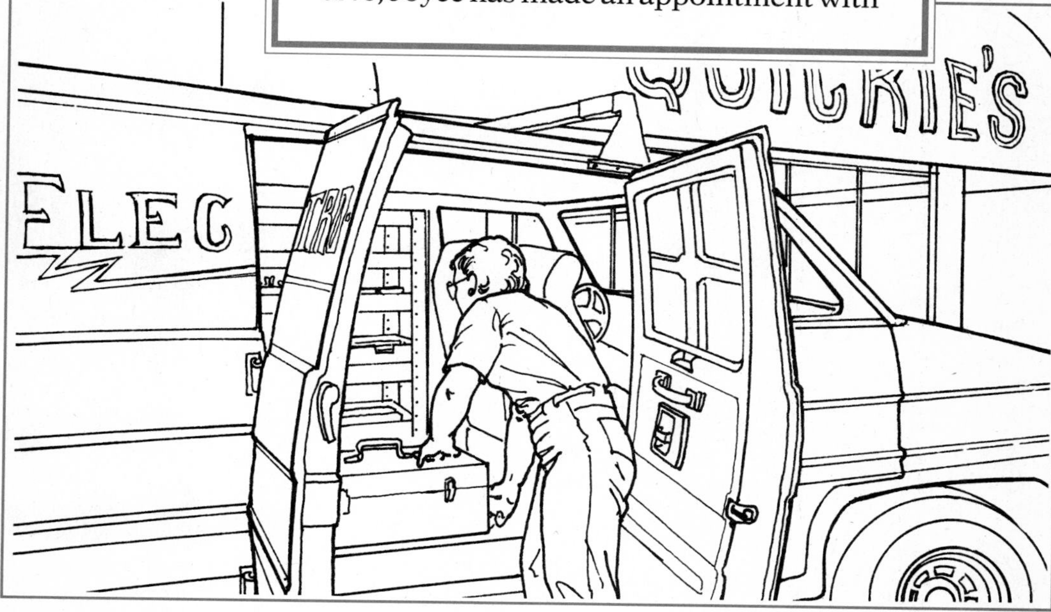

the manager of Quicky's. She has also confirmed her appointment with the electrician.

DEALING WITH CUSTOMERS

Joyce has had some training at Electroserve in dealing with customers. Good manners are always important. In addition, workers must take the needs of the customer into account.

Joyce has to help an electrician install a new menu board, but mealtimes at Quicky's can be hectic. If Joyce walked in at noon with the menu board, she would cause a major business disruption.

It always helps to call ahead. Joyce called the manager of Quicky's yesterday. Together they determined that the job would take about 2 ½ hours. They scheduled the appointment for 2:00 p.m. That time avoids lunch hour and, if all goes well, dinner hour too. Figure 8-1 shows a technician

Figure 8-1 A technician using an electronic address book to call for an appointment.

calling a customer to schedule an appointment. The technician is using an electronic address book.

When Joyce walked in, she made sure to greet the workers who would be affected. She explained how she would proceed. Some repair people barge in and start working. They basically shove people out of the way and cause bad feeling.

After Joyce and several workers brought the menu board in, Joyce asked the manager to check it over. Not only is this a polite way to behave, but it also avoids later problems. Suppose the right-hand section of the menu board is in the wrong place. It would be better to know this before installing the board.

Each type of business has its own needs. Some of the basic rules when going to a site are:

1. Always be polite.
2. Schedule appointments whenever possible.
3. Try not to interfere with whatever business is taking place.
4. Respect the opinions of the workers and managers involved.
5. Clean up whatever mess you cause.

HANDLING WORK SAFELY

Joyce will make sure the menu board gets handled correctly and safely. She knows how to hang the board and start the power.

Knowing the Rules of Safety

The electrician will actually install the menu board to its power supply. The electrician follows certain codes. The codes are for safety. They help people avoid fires and electrical failures.

Before installing the actual board, the electrician works on the power supply. The new menu board requires several

Figure 8-2 A well-lighted menu board.

new connections. It is much larger than the old board. It also has several new electrical connections to handle the new lighting. Figure 8-2 shows a well-lighted menu board.

The electrician has to do some rewiring. His first step is to turn off the power to the menu board. He goes to the circuit breaker box in the back of the store. Figure 8-3 shows an electrician working on a circuit breaker box. He locates the switch labeled "Menu Board" and turns it off. This cuts off the power to the menu board.

The electrician also knows that the power is off because he sees that the light on the old menu board is now off. Turning the power off can save the electrician's life. It is dangerous to try to put in new connections with **live** wires.

Figure 8-3 An electrician working on a circuit breaker box.

Figure 8-4 shows an electrician working on wires after the power supply has been turned off.

Making Doubly Sure

Even though the power is off, the electrician will use tools with rubber handles. Rubber does not conduct electricity. So, just to be doubly sure, the electrician always protects himself from electricity being transferred to his hands. The electrician safely disconnects the old menu board. Next, he prepares the wiring for the new board according to plans. He has already checked the plans carefully before going ahead.

Afterward, the menu board is lifted into place. Joyce makes some temporary attachments to hold it in place while the electrician works. When he completes the new connections, Joyce attaches the board firmly.

Figure 8-4 An electrician working on wires after the power supply has been turned off.

Testing Products On-Site

Before the menu board left Electroserve, the testing department checked all its functions. Now that the menu board is in place, Joyce will do an on-site check.

Knowing What to Look For

Before leaving Electroserve, Joyce spoke to the service manager. She received a function checklist for on-site testing. Figure 8-5 shows an on-site function checklist. She also discussed how she would do the job and any problems the manager thought might come up.

Many things could have gone wrong since Electroserve's tests. Something could have come loose while moving the board. Something in the store's wiring may affect the way the board works.

ON-SITE FUNCTION CHECKLIST

Customer: ____________________ Date: ____________

Technician: ____________________

Item: ____________________

Serial no: ____________________

1. Power turns on and off and switch works easily__________
2. Lights go on as specified in designer's checklist__________
3. Customer uses item to their own satisfaction__________
4. Technician tries tests for on-site specified by designer__________
5. Technician gets approval from customer__________

I agree that the above item performed satisfactorily after installation.

____

Customer

Electroserve, Inc. warranties all our products under normal use. This checklist is our notification that our technician has successfully installed and checked your item to your satisfaction.

Figure 8-5 An on-site function checklist.

Joyce turns on the board. It works. But now, she must see if it will perform its specialized tasks. Quicky's wants the menu board to darken certain sections when that particular part of the menu will not be served. Joyce tries darkening the breakfast menu with a switch marked for the purpose.

Fixing a Problem

What happens next is troubling. The breakfast menu stays lit and the dinner menu darkens. After trying the other switches, Joyce realizes that the switches have been connected incorrectly. However, Joyce solves this problem by relabeling the switches. She retests the board. It works fine.

Doing Proper Paperwork

The electrician leaves. But Joyce has one more job to do before she leaves. She must fill out Electroserve's paperwork.

WORK ORDER FORM

Customer: ____________________ Date: ______________

Problem: ____

____

____

Technician: ____________________

Work done: ____

____

____

Time in: ______________ Time out: ____________

Parts used: ____

____

____

____

I agree that the above work was performed to my satisfaction.

____

Customer

Figure 8-6 A work order form.

DAILY SCHEDULE

Technician: ____________________ Date: ______________

1. Visit Quicky's #14 and install new register pad for drive-thru. Appointment is at 10:00 and work should take 1 ¼ hours.
   Time in: ____________ Time out: ____________
2. Visit Quicky's #27 and repair drive-thru speaker. Can go there at 11:30 and work should take ½ hour. You can work on speaker while drive-thru is still operational and not affect business.
   Time in: ____________ Time out: ____________
3. Return to plant for 1:00 service department meeting.
4. Visit Jambalaya's and install new safety light.
   Time in: ____________ Time out: ____________
5. Visit Quicky's 18 and test out their menu board. Report to supervisor about complaints.
   Time in: ____________ Time out: ____________
6. Return to plant.

Figure 8-7 A daily schedule for a technician.

For every job on-site, the technician fills out a work order form like the one shown in Figure 8-6. Joyce must list exactly what she installed. She also lists the number of workers from Electroserve involved in the installation. In this case, there were two. She puts down the hours worked.

The price of the menu board does not include installation. The customer will receive a bill for that service. The bill is based on Joyce's filled-out form. Joyce had the manager sign the form. She made sure that the manager read the work and hours. The signature helps avoid any disputes.

Joyce returns in the van to Electroserve. She hands in the completed form. She also reports to her supervisor. She gets her schedule for tomorrow. Figure 8-7 shows a typical day's schedule for a service technician.

BILL FOR SERVICE

Customer: Quicky's #32 Date: July 15, 199X

Problem: Menu Board lights a fading.

Technician: José Velez

Work done: Checked board and found short circuit. Replaced 2 x 24 wires; cleaned grease off back of board; tested all functions

Time in: 9:30 Time out: 11:30

Parts used: 2 x 24 wires

Technician's time (includes ½-hour driving time)
2 ½ hours at $35.00 per hour
Total Due: $87.50

Please remit within 30 days.

I agree that the above work was performed to my satisfaction.

____
Customer

Figure 8-8 A bill for service time.

At the end of every day, the service supervisor sends the bills to the billing department. The billing department sends separate bills to customers for service or installation. Figure 8-8 shows a bill for service.

CHAPTER 9

SAFETY IN ELECTRICITY AND ELECTRONICS

At the end of the day, you return to Electroserve where you get to ask questions and talk about safety. You go back to the manufacturing area so that Benjamin Orleans can discuss safety.

Safety plays a major role in all industries. But electrical safety avoids deadly accidents.

Electrical accidents can result in death. They can also result in serious burns. Using equipment related to electricity and electronics also requires attention to safety.

UNDERSTANDING ELECTRICITY

The first step in handling electricity is to understand it. Safety involves an knowing how electricity is transmitted. If a worker knows how electricity works, then he or she will know how to avoid getting hurt.

The Power of Electricity

Electricity is energy or power resulting from the movement of electrons. One of the ways you can see electricity is in lightning storms. A bolt of lightning is a display of electricity. Figure 9-1 shows a display of lightning.

Figure 9-1 A display of lightning.

Lightning has tremendous power. It can split a huge tree right in half. Sometimes, lightning runs along the ground. Such lightning, called ground lightning, can actually burn a pathway in the ground. Figure 9-2 shows lightning striking open ground on a prairie.

Figure 9-2 Lightning hitting the ground on a prairie.

Like lightning, electricity has tremendous power. Sometimes, it has more power than the human body can take. A bolt of lightning can kill a human being in an instant. An electric **shock** can also kill a person in the same way.

Static Electricity

Not all electrical charges pose a danger. Sometimes, particularly in cold weather, you may get a small shock from touching a fabric. This **static electricity** does not injure you. It just reminds you that electricity is all around us.

HANDLING ELECTRICITY

If you live in a country area, you most likely know that a downed power line can kill. Figure 9-3 shows power lines downed after a tornado.

Figure 9-3 Power lines downed after a tornado.

As a child, you were told by adults to keep away from outlets. This and other warnings may have taught you to respect electricity.

Electricity is the major source for power. Someone has to handle it. Service technicians need to keep the power flowing. Repair technicians need to fix machines. Manufacturers need to build electrical and electronic products.

Benjamin tells you, "The first rule in handling electricity is to avoid having the charge hit you. Electricity can only move through or along conductors. The technician can use nonconductors to block the electricity."

Benjamin shows you some of the tools used by Electroserve's technicians and electricians. You notice that all the tools have a black coating on the handles. Benjamin explains, "The tools consist of metal parts for strength. The metal conducts electricity. The plastic and rubber handles that you see do not conduct electricity. If the workers hold the tools by the handles and do not touch the metal, no electricity will get through to them." Figure 9-4 shows several tools with insulated handles.

"The tools can't always protect you, Benjamin continues. Sometimes, electricity can **arc,** or jump over, the handle and

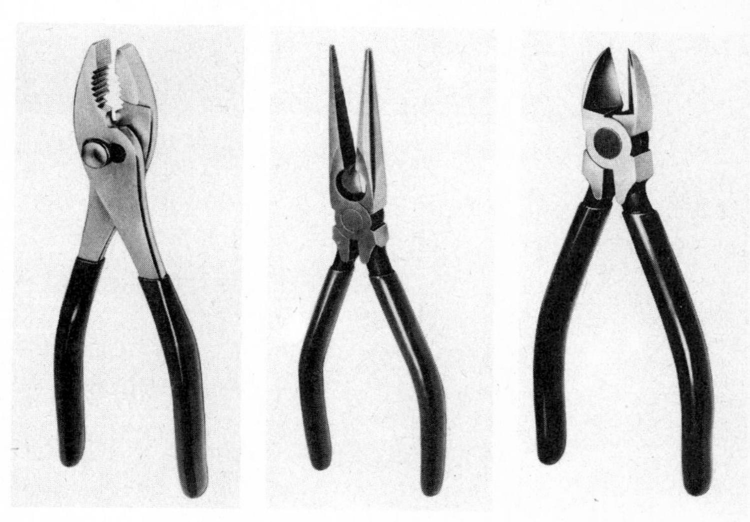

Figure 9-4 Tools with insulated handles.

Figure 9-5 Wires hooked to a testing meter.

get to the worker. This can happen when a tool is used improperly. Sometimes even when a device is turned off, electricity remains inside and can cause injury. Trained technicians know how to **discharge,** or release, electricity in such cases."

Water conducts electricity. If you were to take a radio into a bathtub, you could be **electrocuted,** or killed by electricity. If there is water around, it is best not to touch anything electrical.

Figure 9-6 A technician wearing safe clothing.

When using meters, technicians handle the connections carefully. Most meters have wires that must be connected to current to get a reading. Figure 9-5 shows wires hooked up for reading a meter. Once the reading is made, the way the technician unhooks the wires can mean the difference between a simple procedure and certain injury. The current must be disconnected before unhooking the wires.

Clothing can conduct electricity. A technician should not wear loose, hanging clothes that may get caught up in wires. Also, rubber-soled shoes and rubber gloves can prevent the flow of electricity in certain situations. Figure 9-6 shows a technician wearing safe clothing.

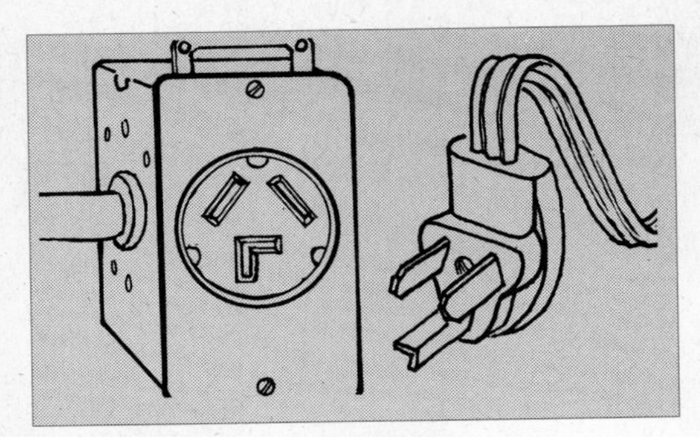

Figure 9-7 A grounded outlet.

Most machines need to be **grounded** when connected to electricity. A ground is a common type of return circuit that can prevent shocks. Figure 9-7 shows a grounded wire being connected to an outlet.

Your class has learned about the tremendous power of electricity. You have also learned about safety. Electrical and electronics technicians are the backbone of perhaps the most important industry in the world—both now and in the future.

GLOSSARY

AC The abbreviation for *alternating current.*

alternating current A type of current flow in which the electrons first move in one direction and then move in the reverse direction.

ammeter A meter that measures current in amps.

amp An abbreviation for *ampere.*

ampere A unit of measure. An ampere is the measure of the intensity of current. Each ampere represents 6,250,000,000,000,000,000 electrons flowing through one point at one time.

analog Represented by measurable quantities. An analog watch has numbers or marks equally spaced to represent the passage of time.

arc To cross over a gap, as in a streak of electricity. As a safety measure, any equipment likely to arc should be disconnected and discharged.

atom The smallest unit of an element. The atom has a nucleus, or central part, surrounded by electrons. The nucleus contains protons.

bit A single unit in computer language. A bit is either a zero or a one.

byte A group of eight bits. A byte represents one piece of information stored in a computer.

cable A group of wires twisted together.

capacitor A device with two terminals that has the capacity to collect and store electric charges.

charge To make something electrically positive or negative. Also, the property of something that allows it to have a positive or negative quality.

chip See *microchip.*

circuit A path through which electrical current flows.

circuit diagram A drawing that shows the circuit in a component or a device. The drawing shows symbols that indicate various characteristics of the circuit, for instance, how it will flow.

component A part of the whole. Two examples of electronic components are a transistor or a diode.

computer An electronic device that processes and stores information. Computers can do arithmetic at high speeds. Computer data is stored in digital form, and the computer can process the data at high speeds, making decisions about and analyzing the data.

conduct To allow electricity to flow through.

conductor A substance or material that provides an easy path for a flow of energy, such as electricity.

current A flow of electric charge.

data Information in numerical form that is stored or processed in a computer.

DC The abbreviation for *direct current*.

digit A single numerical symbol. In computers, digits are usually either zero or one.

digital Representing information as a series of numbers. A digital watch shows the time in a series of numbers, rather than by the position of hands on a dial.

diode A device, such as a semiconductor, which allows current to flow in one direction only.

direct current A type of electric current in which the electrons flow in one direction only.

discharge To remove the stored electric charge from, as in a battery.

electric See *electrical.*

electric charge See *charge.*

electrical Of or from electricity.

electrical circuit See *circuit.*

electricity A form of energy that comes from the movement of electrons and the interactions of electric charges.

electrocute To kill by electricity.

electron A particle inside an atom, which has a negative electric charge. Electrons surround the nucleus, or central part, of the atom.

electronic Of or involved in electronics. Also, having to do with electrons.

electronics The field of technology concerned with developing devices that derive their power from the flow of electrons in active components, such as transistors.

energy Power that can be used to do work. For example, electrical energy can toast bread in a toaster.

flow To move or proceed continuously. Also, a continuous, steady movement. Electrical flow is the steady movement of a stream of electrons.

free electron An electron that is not trapped by the orbit of the nucleus in an atom. Free electrons can move between atoms. They form the basis of electricity.

ground An electrical conductor attached to the earth. Also, to connect electricity to such a conductor. A grounded outlet is considered safe.

hardware The physical parts of a computer, including the monitor, keyboard, central processing unit, and printer.

horsepower A unit of power equal to 745.7 watts. Horsepower is usually used to describe the power of engines.

housing The outside shell of a machine or device that serves as a cover. The housing also holds the interior parts in place either because the housing has slots and openings or because parts are attached to it.

I The symbol for *current*.

IC The abbreviation for *integrated circuit*.

insulator A material that will not conduct electricity. Rubber and some plastics are good insulators.

integrated circuit An electronic circuit package that puts a number of electronic devices together. Examples of such devices are resistors, transistors, diodes, and capacitors.

kit A setup of parts that when assembled complete a machine or a part of a machine.

memory The part of a computer where data can be stored for later use.

microchip An extremely small piece of material, such as silicon, upon which a microcircuit is placed.

microcircuit An extremely small integrated circuit.

monitor The televisionlike viewing device on which you can see data from a computer.

multimeter A measuring or testing device that serves all the functions of an ammeter, an ohmmeter, and a voltmeter.

negative Having an electric charge that is the same as that of an electron so that electrons are repelled. A minus sign (-) is used to indicate a negative charge.

neutron A unit inside the nucleus, or central part, of an atom. It is neutral and does not have either a negative or a positive charge.

ohm A unit of measurement that equals one unit of the resistance of a conductor to the flow of electrical current.

ohmmeter A measuring and testing device that measures ohms.

oscilloscope An electronic instrument that displays the wave action of a current on a screen. Oscilloscopes have a cathode-ray tube for the display.

path The route along which electrical current moves.

positive Having an electric charge that can attract electrons because the charge is the opposite of the negative charge within an electron. A plus sign (+) indicates a positive charge.

potential energy Energy that is available or stored in a place. Energy that is not from the movement of electrical current. A good example of potential energy is a spring that is tightly coiled. When let go, the stored energy releases the power of the spring.

printed circuit board A thin piece of material, such as fiberglass, on which conductive material has been placed into etched pathways. The pathways are for the circuitry.

program A set of instructions used to run a computer or one specific set of tasks on a computer.

proton A particle within an atom that has positive electrical charge.

resistance The quality of a circuit that prevents electrons from flowing through it.

resistor A device that prevents the flow of electrical current.

semiconductor A material that has a resistance somewhere between that of a good conductor and a good insulator.

shock The startling feeling caused by electric current going through the body or any one of its parts.

short circuit A path across a source. The path provides zero resistance to the electrical current so that the current goes around the devices in a circuit. A short circuit usually causes equipment failure.

silicon A material, often used in transistors or as wafers in boards, that allows for the easy flow of electricity.

software Information, usually put in manuals and on disks, that contains the instructions, programs, and symbols necessary to run a computer or a specific set of tasks on a computer.

solder To join two metal pieces together with a molten combination of tin and lead. Also, the material itself. When cool, the solder hardens and holds the connection fast.

static electricity Electricity accumulating in one place as opposed to electricity moving in a current.

strand One of the wires wrapped together to form flexible wire.

superconductor Any of various materials that have virtually zero resistance. Superconductors are playing an important role in the development of new electronic products.

tester A device or machine that measures or tests something, such as the qualities of a circuit.

testing device See *tester*.

transistor A semiconductor device with three terminals that is used in increasing, changing, or switching electrical action.

upgrade To change for the better, for example, to upgrade an electronic device by adding a new printed circuit board so that it can perform more functions.

V The abbreviation for *volt*.

volt A unit of measure equal to the electric potential. It is measured between the ends of a conductor with the resistance of one ohm and with a current of one ampere flowing through it.

voltage The amount of electric potential or volts.

voltmeter A testing device that measures the voltage.

watt A unit of electric power, as in a machine or a device such as a bulb.

INDEX